W9-BZA-170

University of California Press / Monterey Bay Aquarium
Series in Marine Conservation, 2

"Healthy oceans are critical to the future of all life on Earth, yet by and large the underwater world remains hidden to us, unknown and mysterious. The mission of the Monterey Bay Aquarium is to inspire conservation of the oceans, and this series of books is intended to further that goal. By helping people discover their connection with the natural world, we hope to foster a lifelong commitment to learning about and caring for the oceans on which all life depends."

—Julie Packard
Executive Director
Monterey Bay Aquarium

A Living Bay

A Living Bay

The Underwater World of Monterey Bay

LOVELL LANGSTROTH and LIBBY LANGSTROTH

Todd Newberry, Editorial Associate

UNIVERSITY OF CALIFORNIA PRESS

Berkeley Los Angeles London

MONTEREY BAY AQUARIUM

University of California Press
Berkeley and Los Angeles, California

University of California Press, Ltd.
London, England

MONTEREY BAY AQUARIUM®

Library of Congress Cataloging-in-Publication Data

Langstroth, Lovell.
A living bay : the underwater world of Monterey Bay / Lovell Langstroth and Libby Langstroth ; Todd Newberry, editorial associate.
p. cm. — (University of California Press / Monterey Bay Aquarium series in marine conservation ; 2)
Includes bibliographical references (p.) and index.
ISBN 0-520-21686-5 (cloth : alk. paper) —
ISBN 0-520-22149-4 (pbk. : alk. paper)
1. Marine ecology—California—Monterey Bay. 2. Marine biology—California—Monterey Bay. I. Langstroth, Libby, 1921– II. Newberry, Todd. III. Monterey Bay Aquarium. IV. Title. V. Series.
QH105.C2 L36 2000
577.7'432—dc21 00-022014
CIP

Manufactured in Singapore
09 08 07 06 05 04 03 02 01 00
10 9 8 7 6 5 4 3 2 1

The paper used in this publication meets the minimum requirements of ANSI/NISO Z39.48-1992 (R 1997) (*Permanence of Paper*). ♾

The photo on page ii is of giant kelp; that on page v, of a nudibranch *(Flabellina iodinea)*. The map on page xvi and figures 1–3 were drawn by Bill Nelson. Figures 4–14 were drawn by Freya Sommer.

To Chuck Baxter, our mentor,
who introduced us to the organisms of Monterey Bay

CONTENTS

FIGURES

PREFACE

We are going to tell you about how some organisms in Monterey Bay, California, lead their lives. To this end we have combined photographs with commentary about each image. In this region, northern and southern species of plants and animals overlap in their respective ranges, making the bay's floral and faunal diversity truly unparalleled in the eastern North Pacific. For example, some 80 percent of the seaweeds known from Baja California to Alaska occur in Monterey Bay, and the bay's intertidal invertebrate diversity (and probably its subtidal diversity as well) is comparably awesome. Thus, for our purposes of inquiry, the Monterey Bay can stand for the whole central Pacific Coast. The bay encompasses a surface area of almost one hundred square miles of oceanic water and nutrient-rich inshore waters, and its grandest underwater feature, the Monterey Bay Canyon, has a depth and profile comparable to those of the Grand Canyon of the Colorado River.

Ten major marine research centers encircle the bay, making it one of the world's most intensively studied large natural areas, from its shoreline to the outer bay waters and down to the depths of the submarine canyon. The internationally acclaimed Monterey Bay Aquarium faithfully recreates the bay's habitats. And the waters off the California coast from Cambria in the south to the Marin headlands in the north have been declared a National Marine Sanctuary.

We are the fortunate beneficiaries of these resources. Together, cameras in hand, we have made many hundreds of dives in the bay's waters, observing and photographing organisms in their natural habitats. We have explored the seashore at low tides. In our laboratory, we have photographed with magnification many tiny creatures that have revealed unexpected secrets. Trawls and specialized laboratories have provided us with extraordinary organisms recovered from the depths. Specialists at universities have freely made themselves available for consultation, and outstanding marine libraries have opened their doors to us as we surveyed the enormous literature.

We have arranged the chapters by major habitats, extending outward and downward from the shore to the depths of the Monterey Canyon. Within chapters topics are organized not by phylum but rather by location and association within the habitat. Although many organisms have great beauty to be sure, we are concerned principally with function. Notwithstanding their diversity, the welter of organisms in this book are linked by a few recurrent themes:

Sexual interactions We focus on the ways organisms solve various problems that attend reproduction, such as finding a mate, synchronizing spawning, and achieving fertilization.

Predator-prey interactions We depict predators caught in the act, and prey relying on visual camouflage or mimicry to confuse the enemy—moments when one creature's meal is another's life.

Chemical interactions Chemical camouflage, communication with pheromones—such functions are pervasive in the natural world. Many molecules serve one purpose in invertebrates and another in humans: a protein defends mussels and sponges from toxins; it also defeats chemotherapy in human tumors. We will explore such fascinating relationships throughout this book.

Now it is time for us to reach out to tidepoolers, to divers, to aquarium visitors, to armchair naturalists, to students, through the medium of our photos combined with text, and ignite a shared passion for the organisms we depict. So settle into the spirit of the book, a spirit of fascination with the natural history beneath the surface of Monterey Bay.

ACKNOWLEDGMENTS

We are indebted to many people who have helped us since the inception of *A Living Bay.* In particular we owe thanks to our Editorial Associate, Todd Newberry. He has given us much needed encouragement and guidance from our rather tentative beginnings right through to the finished product. He has read and reread the text, helped with organization and style (including issuing stern admonitions to use the active voice), and struggled to keep two headstrong authors from straying too far afield. Scarcely a page fails to bear his graceful imprint.

It was Dan Gotshall who built a fire under us, inspiring us to undertake this book in the first place. Both Chuck Baxter and John Pearse read the entire manuscript in early incarnations, and portions of the text-in-progress were read by David Epel, Mike Foster, Leon Hallacher, Melissa Kaufman, Robert Lea, Welton Lee, James Nybakken, Pam Roe, Stuart Thompson, and Russel L. Zimmer. All authorities in their respective fields, they made wonderfully helpful suggestions and uncovered errors. Of course, we take sole responsibility for the finished product, and we invite our readers to bring to our attention any problems that will inevitably have crept into a work of this complexity.

Joe Wible, Hopkins Marine Station librarian, tracked down many obscure and out-of-print references. We benefited from Freya Sommer's knowledge of cnidarian life cycles; she contributed the unique photographs of their life history stages, and she drew the meticulously executed line drawings that accompany discussions of various organisms' life cycles. Greg Jensen, Patsy McLaughlin, Peter Slattery, Dorothy and John Soule, and Mary Wicksten are among those who helped with problems of identification. Gil Van Dykhuizen introduced us to the mysteries of the Monterey Bay Aquarium's deep sea laboratory. Lynn and Keith Chase helped with tedious organizational details of the photos. It's heartwarming to have had so many contribute so freely to our effort.

Anne Canright, the copyeditor, was our initial, all-important contact with the Press; the text has benefited from her vast experience. Additionally, we can only marvel at the Press, that mysterious entity that churns out such outstanding publications. Acquisitions editor Doris Kretschmer, project editor Rose Vekony, designer Barbara Jellow, production coordinator Laura Paulini, and others on the team worked magic, integrating many complex elements into a beautifully presented, unified whole. It has been a privilege working with them. Finally, we are indebted to the David and Lucile Packard Foundation, whose generous grant facilitated this costly undertaking.

All the photographs were taken by the authors unless otherwise indicated.

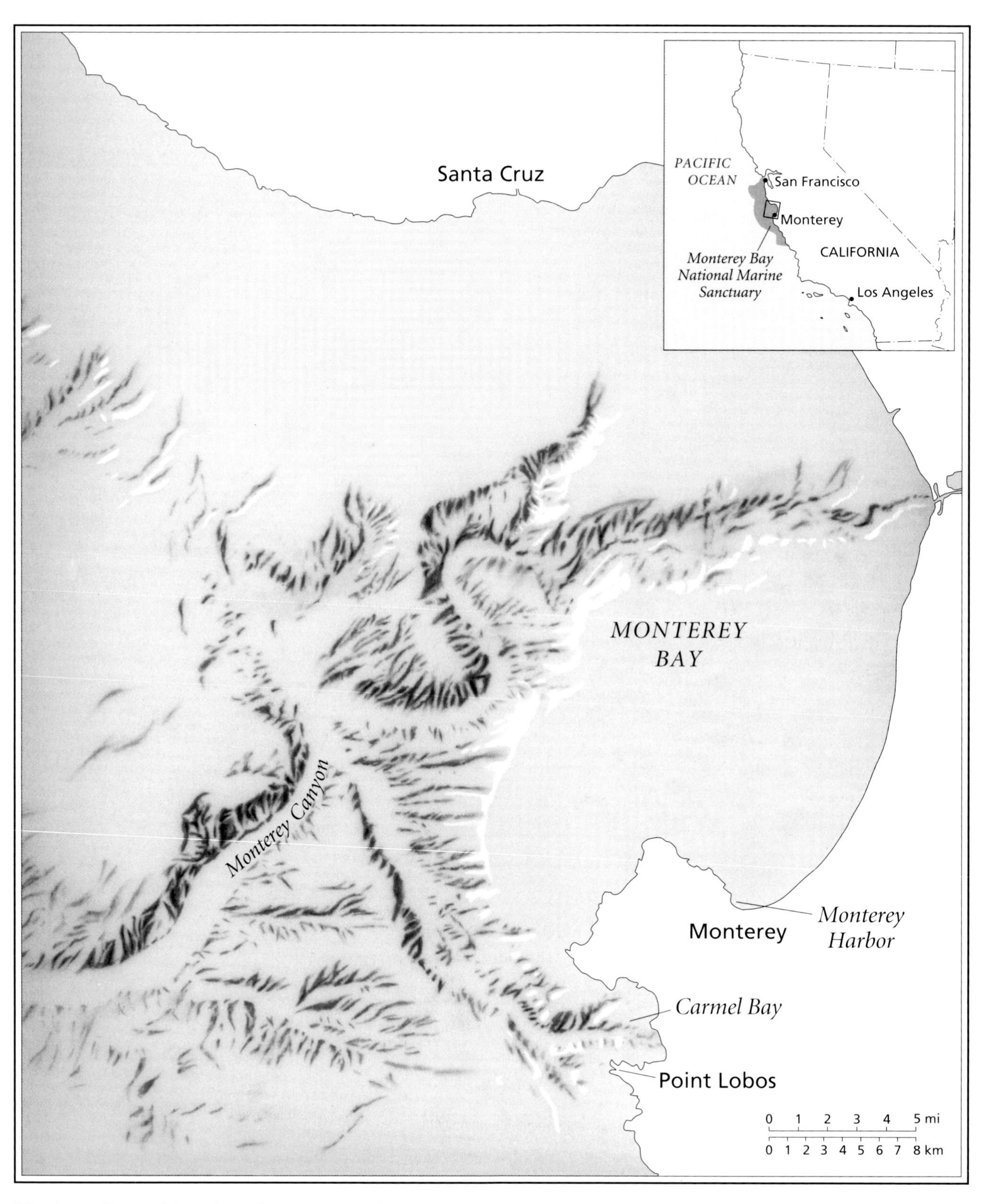

Monterey Bay and its submarine canyon system.

INTRODUCTION

From its beaches and intertidal rocks to the bottom of its submarine canyon, Monterey Bay is one of the richest and most diverse environments in the world: rich in that it supports huge numbers of organisms, and diverse in that a wide range of physical habitats support multitudes of species in many major groupings. Far more major groups of animals are represented here than in a tropical rain forest. And of course these organisms require widely different adaptations to confront widely different physical conditions. On intertidal rocks they are exposed to summer heat and, when the tide is out, to desiccation, whereas in winter they encounter freezing temperatures and deluges of fresh water from seasonal rains. On the canyon floor, organisms are subjected to enormously high pressure, temperatures approaching freezing, a complete absence of locally derived oxygen and organic products from photosynthesis, and exposure to the toxic hydrogen sulfide. We will tell you how some of these organisms live and how they fit into and in many ways contribute to this astonishingly bountiful environment.

Before we begin our journey through these habitats we would like to review some basic ideas about marine organisms, especially their life cycles, and acquaint you with some useful vocabulary in the process. Many readers will be familiar with this information, but some of it may be new to you, and key ideas can always bear restatement. Note that in the discussions of individual organisms we assume readers' familiarity with these terms and concepts, whether from previous exposure or from having read this introduction, but we also provide a glossary for reference. In addition, an appendix lists the major groups, or phyla, of marine animals, with discussion of their major characteristics.

Upwelling, Seasonal Changes, and the Food Web

The richness and diversity of the marine life of California's central coast are due in part to the area's high productivity, which depends on the upwelling of coastal waters. In the spring and summer strong northwest winds drive the surface waters along the coast. This water slides right of the prevailing winds, shifted by the rotation of the earth (Coriolis effect). As the surface water moves southward and hence offshore, it is replaced by cold water welling up from the depths. This deep water contains nitrates and phosphates—vital nutrients—derived from sinking excretions and decomposing organic matter. Concurrently, the days grow longer, the sun is higher in the sky, and light penetrates the water longer and more deeply. The combination of increased light and nutrients works magic: algae large and small undergo a burst of growth. Giant kelp, the dominant alga of the kelp forest in Monterey Bay, may grow over a foot a day during this time. Microscopic drifting algae, or phytoplankton, multiply in great blooms, and the ocean becomes a turbid green. Visibility may be reduced to three feet; as divers, we can scarcely see our own fins.

Brought to the sunlit surface, the upwelling nutrients and blooming phytoplankton support prolific food chains. Tiny drifting animals, or zooplankton, multiply rapidly, feeding on the phytoplankton and on each other. Especially now, but all year long to some degree, suspension feeders devour phytoplankton, zooplankton, and drifting bits of detritus (particulate organic matter). Herbivores graze on fleshy algae; deposit feeders take in sedimental detritus; carnivores eat herbivores and each other; and various omnivores exploit food of all sorts. When we confront specific animals within these broad categories of diet and feeding strategy, we find that neat, abstract, linear energy-flow diagrams like food chains become complicated, branching and converging webs of predator-prey interactions that dominate the lives of real organisms.

Fall ushers in the oceanic period. Northwest winds die out, followed by the relaxation of upwelling and the flow of warmer, nutrient-poor, clear water inshore as upwelled water begins to sink. It is then that spectacular jellyfish, salps, and other members of the offshore planktonic community may enter the bay and present themselves to our cameras. It is also then that "red tides" of dinoflagellates may signal the presence of dangerous toxins in the food web.

Finally, winter storms bring long-period swells, and the accompanying underwater surge rips out kelp holdfasts and tears away the kelp's surface canopy. As pounding surf attacks the shoreline, attached organisms must withstand the enormous pull of surging water; the impact of floating logs clears away patches of rocky shore communities, creating new opportunities for colonization; and beaches almost disappear as sand is swept away, exposing the underlying rocks. Monterey Bay's seasonal changes could scarcely be more dramatic.

Plants and Animals

Many marine animals at first glance appear to us as "plantlike." Sea anemones, with their petal-like tentacles, do resemble flower blossoms. Yet Aristotle, in the fourth century B.C., observed sea anemones ingesting other animals: "If any little fish comes up against it it clings to it; . . . and it feeds upon sea-urchins and scallops" (Thompson 1910, 531a). Finally in the early eighteenth century, the French surgeon J. A. Peyssonnel ended centuries of uncertainty by experimentally showing that the polyps of precious red coral, essentially a colony of anemones, feed themselves—that corals are, in fact, animals. In recent decades, the old either/or distinction between plants and animals has given way to a six-kingdom—or more—classification of living things. Two of these kingdoms are the Animalia and the Plantae. Seaweeds (attached algae) have been variously classified; we follow current practice and place them not among the Plantae but in the kingdom Protista (Singer 1997), which also includes single-cell protozoa and most phytoplankton. Plants, which had their evolutionary origin in the green algae, differ from algae in that they develop from an embryo protected by maternal tissue, such as a seed coat or fruit.

We are concerned in this book principally with algal seaweeds and animals. Both algae and animals respire, burning carbohydrates to release energy and, as a by-product, carbon dioxide. They differ, however, in one fundamental way. Algae (and plants) are "primary producers": they produce new organic compounds by photosynthesis, using the sun's energy to form simple sugars from even simpler carbon dioxide and water, with the help of a chemical catalyst, chlorophyll. This process supplies their food and material for growth. Animals, in contrast, cannot produce their own food. They are consumers: they must obtain their energy from food that plants or algae (or some bacteria) already have produced or that other animals have incorporated through ingestion.

The Cell

It was as recently as 1839 that the botanist Matthias Jacob Schleiden and the zoologist Theodor Schwann, after comparing microscopic sections from plants and animals and recognizing many similarities, proposed that all living organisms are composed of cells. How far we have come since then! It is today general knowledge that most cells of eukaryotic organisms—those with a nucleus—have two sets of chromosomes within their nuclei, each a complete set of genetic instructions from the parents. They are diploid. Eukaryotes include all living organisms except bacteria and cyanobacteria (blue-green algae). During an organism's life most of its cells replicate their chromosomes by means of mitosis, or cell division: following division, the daughter cells have in their nuclei copies of the same two sets of chromosomes that the parent cells did. In contrast, the chromosomes of eggs and sperm (gametes) are formed by meiosis, or reduction division. In meiosis, a diploid cell's two sets of chromosomes are halved, and at the same time their sequences of genes are rearranged as chromosomes exchange bits and pieces. The result is a genetic diversity of haploid gametes, or male and female germ cells having only a single set of chromosomes per nucleus. The subsequent union of egg and sperm through fertilization results in a new diploid individual, the zygote, with an unpredictable mix of parental genes.

Seaweeds

Although algae are, strictly speaking, evolutionarily primitive plants, coastal seaweeds, the larger attached algae, differ markedly from terrestrial plants. They do not have differentiated true roots, leaves, or stems or the internal specialized water- and food-conducting (translocating) structures of vascular plants. Rather, the algal body, known as the thallus, consists of an anchoring holdfast, a stemlike stipe, and filaments or leaflike blades. There is a low diversity of cells. Nutrients are absorbed directly from the water through the entire thallus. The seaweeds traditionally are classified in three major divisions based on the color of their photosyn-

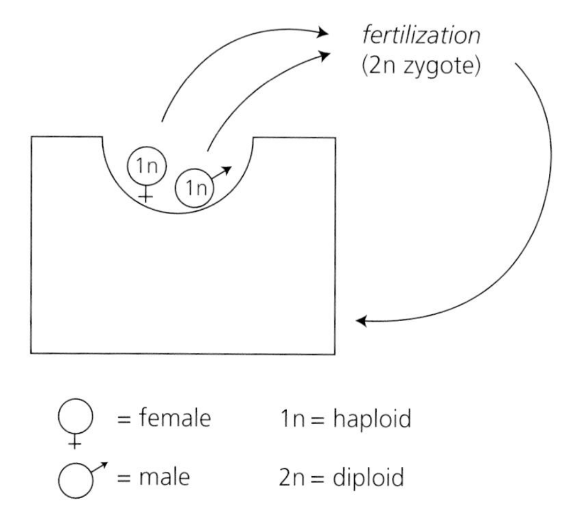

Fig. 1. Some brown algal life cycles
e.g., *Fucus gardneri*

thetic pigments: the green, the brown, and the red algae. They all have chlorophyll *a*, a green pigment. In the brown and red algae, accessory pigments mask the underlying chlorophyll's green color.

Interestingly, an enzyme that speeds photosynthesis as a whole (RuBisCO, the world's most abundant protein) is two or three times more efficient in red algae than it is in higher plants, including crop plants. What if one could transfer the enzyme-producing genes from red algae to these plants? It could produce a new agricultural revolution. This complex task, "the most fundamental genetic alteration that humankind has ever tried in any organism," is currently being attempted (Mann 1999).

The life cycles of different kinds of algae vary markedly. Some brown algae (e.g., *Fucus* and *Cystoseira*) exist only as a diploid thallus that gives rise to gametes by meiosis (Fig. 1). After fertilization, the resulting zygotes grow directly into new thalli.

Most other brown algae (e.g., *Macrocystis*) and all the green algae (e.g., *Derbesia*) have two alternating phases in their life cycles (Fig. 2): a sporophyte that produces spores and a gametophyte that produces gametes (the suffix *-phyte* is derived from the Greek word *phyton*, which means plant). In one phase of the life cycle, the diploid sporophyte meiotically produces dispersive spores that germinate into haploid gametophytes. In the other phase, the gametophytes mitotically produce eggs or sperm. Fertilization of the eggs then leads to another diploid sporophyte generation.

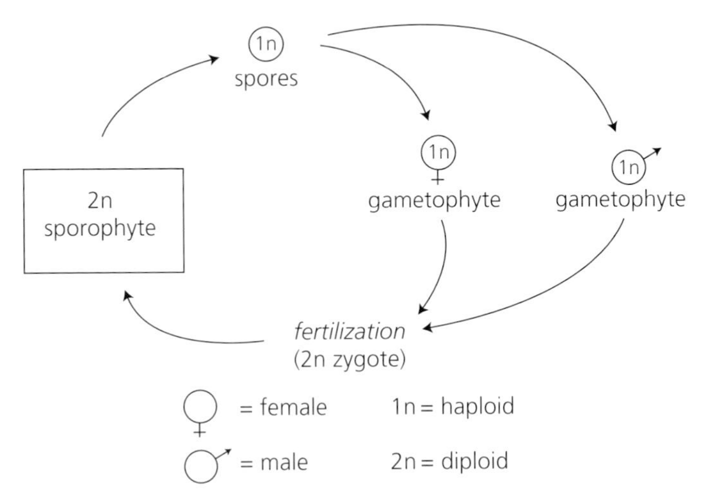

Fig. 2. Green and most brown algal life cycles
e.g., *Derbesia marina, Macrocystis pyrifera*

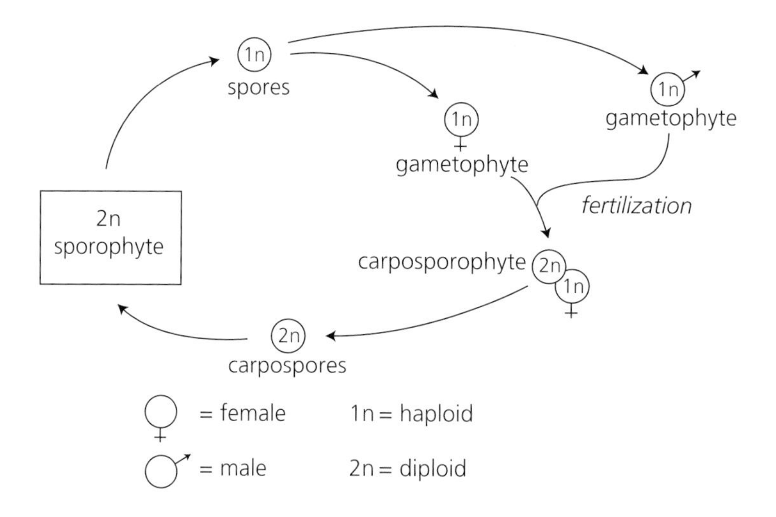

Fig. 3. Red algal life cycles
e.g., *Mastocarpus papillatus, Fauchea laciniata*

Red algal life cycles (e.g., *Mastocarpus, Fauchea,* and *Melobesia*) are more complex: they not only have alternating sporophyte and gametophyte phases, but they also have a third phase, a carposporophyte, which forms only on the female gametophyte (Fig. 3). In this phase, sperm fertilize eggs on the female gametophyte, but the zygotes are not released. Instead they undergo further complex development on the female thallus and form another kind of spore, the diploid carpospore. The carpospores are then released and disperse, producing the next "orthodox" sporophyte phase. Thus, to summarize this complex situation, red algae have three sets of reproductive elements: haploid spores from sporophytes, haploid gametes from male and female gametophytes, and diploid carpospores from carposporophytes.

In many algae it is hard to tell a sporophyte blade from a gametophyte blade. Sporophyte and gametophyte phases that are identical in appearance are "isomorphic" (e.g., *Fauchea* and *Melobesia*). But in many algae the sporophyte and gametophyte phases are so different in appearance that it was only through culturing that their identity as a single species was revealed. These algae are "heteromorphic" (e.g., *Macrocystis, Derbesia,* and *Mastocarpus*).

The Invertebrates

The methods of reproduction in invertebrates are not as unfamiliar to us as the life cycles of algae. Nevertheless, many of the developmental strategies of these animals are also quite complex. Invertebrates that spawn by broadcasting may release thousands or even millions of eggs and sperm into the sea, trusting luck—or, more likely, chemical attractants (pheromones)—to bring about fertilization in midwater. While the mothers need not invest much energy apiece in these tiny offspring, the vast numbers of eggs that they produce as a hedge against unsuccessful fertilization and larval calamities certainly take their toll. In other invertebrates, fertilization of the eggs may occur internally as released sperm find and enter the female from the water; in still others, it is achieved by copulation. In these cases, a fertile female may brood only a few eggs, devoting considerable energy to each offspring but thereby enhancing their chances for survival—a very different reproductive strategy from that of broadcast spawners.

Among familiar vertebrates, males and females are usually separate individuals. But many marine invertebrates are hermaphrodites, either simultaneously male and female or sequentially first one sex and then the other. The term derives from the Greek myth of Hermaphroditus, the son of Hermes and Aphrodite, who was adored by the nymph Salmacis; when he bathed in her lake she embraced him and persuaded the gods to join them into one body, both male and female. Various mechanisms exist to prevent self-fertilization in hermaphrodites; cross-fertilization ensures genetic variation. But the real advantage of hermaphroditism is that any encounter is potentially sexual and may also result in the impregnation of both individuals rather than just one.

In many animals, the fertilized egg (zygote) develops directly into a juvenile that is a miniature of the adult form. We humans do this; so do land snails and grasshoppers, for example. Or the zygote may develop into a larval form first that must then undergo metamorphosis, or a change in form, before it assumes the adult appearance, in a process known as indirect development—as in frogs, which develop from tadpoles, and flies, which develop from maggots, to cite but two familiar examples. Sometimes these contrasting patterns of development distinguish rather closely related species, indicating the extent to which every part of the whole life cycle is molded by selective forces.

Larvae that derive the energy to survive to metamorphosis from the yolk of their large egg are "lecithotrophic": they do not feed from the plankton as they drift along. As a rule, these larvae drift only briefly in the plankton and metamorphose within hours or days. Other larvae are born or hatch yolk-free and soon feed from the plankton, drifting along with their food for weeks or even months. These "planktotrophic" larvae often disperse widely—for good or ill—during their planktonic life. In each larva's sojourn in the plankton, of course, it runs the risk of being eaten by midwater predators; the proportion that makes it through to settling is thus likely to be minuscule regardless of larval strategy, brief or lengthy.

If the postlarval stages of the life cycle are benthic (bottom-dwelling), larval settlement and metamorphosis go hand in hand. Tiny as they are, the larvae must "predict" from physical, chemical, and biological information whether the substratum is appropriate for their lifestyle and needs. Many larvae, such as those of various sessile marine worms and barnacles, settle on or near adults of their own kind, apparently homing in on a "chemical essence" of their species. Others may seek out symbiotic or commensal partners or a future food source as their substratum. The same is true of many algae as well, which may grow only on a specific substratum such as another algal species.

Settling larvae not only must choose their sites, they must also attach themselves—no easy task for a pinhead-sized creature at the mercy of strong currents or surf! And then each must defend its site against all the other settling larvae arriving on the scene—and all this on a substratum that is likely to be teeming with already settled animals that feed by filtering the water for . . . larvae! The competition for space and intense predation by benthic suspension feeders combine to suggest that, for larvae, the risks of settling outstrip even the risks of life in the plankton. It is certainly an exceedingly perilous stage in the larval life cycle, and probably one that very few larvae survive.

Sexual and Asexual Reproduction

Even though it is all but universal in algae and animals, the "why" of sex remains unclear. This "why" is obscured by uncertainty even about "the goal of the game": Is it to sustain genes? To sustain lineages of whole organisms? Dawkins (1976) argues engagingly that we are merely "survival machines"—robot vehicles blindly programmed to preserve our "selfish genes." Be that as it may, it is clear that rearranging genes sexually through meisosis and fertilization ensures genetically varied offspring, at least some of which may be able to sustain half a parent's genes into the future amid changing environmental conditions.

Many organisms grow and multiply asexually; by budding or splitting, one body mitotically generates identical genetic copies of itself—a process known as modular growth. Often such organisms have pronounced powers of regeneration. If the newly generated bodies separate from each other, clones of genetically identical individuals result; if they remain organically interconnected, the result is a colony. The clonemates or colonymates may be short lived, but clones or colonies may, by continually reproducing, live for a very long time—years or even centuries. Individuals in a colony can differentiate anatomically to perform different tasks—feeding, defense, nurturing embryos—and thus divide the labor of the whole colony. Organisms living in clones or colonies are far more common and widespread than is generally realized; in the following pages we provide examples from the hydroids, anemones, soft corals, bryozoans, and colonial tunicates living in many different habitats.

The Thirty-five Phyla

The very unfamiliarity of life in the sea merits a pause. Seaweeds, of course, are superficially familiar organisms, even though they differ in many ways from terrestrial plants. But marine animals sometimes look more alien than anything cooked up by Hollywood as extraterrestrial. Why? The answer lies largely in their environment: the sea. Marine organisms in general face fewer physical constraints than their terrestrial counterparts. They are usually surrounded by water, which decreases the need for skeletal support. They are not likely to dry out, they are exposed to fewer temperature extremes, and they encounter little if any UV radiation. On land, the designs that prevail have sequestered the animals' essential organs. In the sea, by contrast, many physiological arrangements function "naked"—external gills instead of lungs and tracheae, broadcast spawning of gametes instead of copulation, mucus feeding nets instead of grab-and-eat feeding, bodies of surpassing delicacy instead of the fuselages and weight-supporting skeletons we find on land. So it should not surprise us that on dry land we regularly encounter only five of the roughly thirty-five great divisions of animal species, or phyla. The other thirty are mostly or wholly marine. They include the phylum Porifera, the sponges, masses of cells that form just a few tissues in amorphous bodies; the phylum Cnidaria, the flowerlike sea anemones, hydroid polyps, and umbrella-like jellyfish medusae, all with stinging cells and a saclike body cavity; the phylum Annelida, the marine segmented worms, motile or in tubes, and angleworms on land; the phylum Arthropoda, the insects and spiders on land and the shrimps, crabs, barnacles, and their relatives in the sea, all with hard exoskeletons but with joints to permit movement; the phylum Mollusca, the mostly shelled snails, clams, chitons, and squid and octopus; the Lophophorates, three phyla of animals that feed with a unique filtering crown of tentacles; the phylum Echinodermata, the seastars, urchins, and their relatives; and evolutionarily closer to home, the phylum Chordata, the relatively unfamiliar tunicates, which show unmistakable chordate traits in their "tadpole" larvae: a dorsal tubular nerve cord, gill slits, and a stiffening rod or notochord (precursor of the spinal cord). Chordates include, of course, the familiar fishes, amphibians, reptiles, birds, and mammals. A more detailed description of the distinguishing traits of organisms in these phyla is given in the appendix—if you've never encountered some of the names just mentioned, it may help your understanding to begin the book with a perusal of this material.

We have presented some basic concepts that should be helpful in interpreting the material that follows—how algae, plants, and animals differ from one another; some basic life cycles of algae and invertebrates; and sexual and asexual reproduction.

The sea is a vast, three-dimensional world with complex food webs embracing bacteria, phytoplankton, zooplankton, and vigorously swimming nekton, or larger animals, from anchovies to whales, that can make headway against strong currents. It is a world where the larval offspring of benthic parents can be cast to the watery habitat the way that, on land, only seeds and pollen are dispersed. It's a world where sessile organisms, unable to pursue prey, prosper from suspended water-borne particles of food. It's a world where drifting molecules profoundly influence an organism's behavior. Marine organisms can exploit entire strategies that are foreclosed to terrestrial organisms. This is one of the reasons they often seem so strange to us, unfamiliar bodies reflecting unfamiliar lives.

Life on land is entirely benthic. We crawl on the bottom of the earth's sea of air. There is almost no aerial plankton—such as pollen for hay-fever sufferers or insects for swallows and bats. Birds in their daily rounds behave like "terrestrial flatfish" as they fly from one benthic resting place to another. In the sea, however, the whole volume of the habitat is available as living space, from the surface waves to the deepest reefs and sands. From passively buoyant or weakly swimming plankton to the robust swimmers of its nekton, the sea boasts ways of midwater life that the atmosphere simply cannot support. A familiar comparison notes that the seas occupy almost three times more geographic expanse than the land. But as a volume that supports life, the seas are fully a thousand times the size of the land's grounded habitats.

In the following chapters we will introduce you to some of the organisms in a stupendous bay, organisms that may at times seem familiar, only to reveal sophisticated and unanticipated behaviors. We hope these organisms will command your attention as they have ours. We hope, too, that you will regard them all, seaweeds and animals alike, with a growing sense of respect and even awe.

1 The Intertidal Zone

The meeting of land and sea presents a special habitat that alternates between these two worlds to the beat of the tides. Exposed to sun, wind, and rain, pounded by surf, and submerged daily in cold water—can you imagine a harsher life? Yet intertidal organisms thrive in these conditions, and in fact many of them even require this ecological mix of terrestrial and marine environments through some or all of their lives. Rocky intertidal habitats undergo the predictable rhythms of tides and seasons, though these may be punctuated unpredictably by trivial to calamitous disturbances. Sandy shores gradually build sloping beaches in calm periods, only to have them reshaped by the first waves of a storm. The intertidal zone is a chaos of microhabitats in flux. Some intertidal organisms have short lives or are transient intertidal residents, whereas others—anemones, as a prime example—may persist for decades in one spot while their world changes around them.

A meeting of land and sea. Whaler's Cove, Point Lobos.

The rocky seashore. Low tide on a calm day.

Intertidal habitats having similar physical settings are strikingly alike throughout the world (Stephenson and Stephenson 1949). Though individual species are different, the communities of organisms on Monterey Peninsula's intertidal rocks are very similar to those of temperate rocky shores elsewhere. If you view a steep rocky seashore at low tide, you can see horizontal bands (zones) of different algal and animal communities extending from high dry rocks down to tidepools. Closely examine our photograph and you will notice a few scattered tufts of dark red algae on high intertidal rocks. Move down the slope, and the cover of algae becomes continuous. Still lower, at this very low tide, glistening surfgrass hangs exposed on the rock, and a shallow-water kelp (*Laminaria dentigera*) bobs in the water. This photo is not detailed enough to reveal the similar succession of invertebrates, but it is there—from periwinkles and a few limpets on the highest rocks down through zones of acorn barnacles, mussels, and stalked barnacles, aggregating anemones, and an increasing diversity of wet-world creatures.

The causes of this zonation are complex. Physical factors such as surf, tides, light and shade, and extremes in temperature and salinity all set upper limits to the ecological ranges of intertidal organisms, whereas biological factors, especially predation, probably set lower, subtidal limits for many (even the same) organisms.

Right: A tidepool at low tide on a calm day.

This tidepool is filled with a mixture of green, red, and brown algae dominated by *Leathesia difformis*, an annual brown alga. The little basin may be deep enough to remain filled with water much of the time, and, like plants in a garden, these seaweeds respond positively when they stay wet.

Compared to intertidal rocks, a sandy beach looks like a desolate place. Clams, crustaceans, and worms find refuge from weather or predatory birds at low tide, or from fish at high tide, by burrowing into the sand. But if we look very closely—and with the aid of a high-power microscope—we discover an enormously rich community in miniature, the meiofauna, consisting of tiny animals and single-celled organisms small enough to live in the water film that adheres to wet sand grains. The existence of this fascinating community has been recognized only in recent decades, and few naturalists have studied it yet on the Pacific Coast.

Through our images we can share a field trip to the seashore. We will examine a variety of algae and invertebrates from the different intertidal zones and discuss their adaptations to this challenging environment. A word of very serious caution, though, should you decide to take your own field trip: the intertidal zone can be as treacherous as it is attractive—no surface is more slippery than wet seaweed. And rogue waves may roll in seemingly out of nowhere. Make sure surf-breaking rocks are between you and the open sea, and always keep one eye on the surf. Never tidepool along the open coast without a buddy. The ocean is big, and we are small; intertidal encounters with the sea tend to be one-sided. Finally, respect environmental regulations. The seashore has evolved as a rough-and-tumble place, but in very particular ways, which have not included human feet and hands. Be a safe visitor and a biologically courteous one too.

Emerita analoga
Phylum Arthropoda / Class Crustacea

Male sand crab, about 15 mm long.
Carmel Beach.

Look for the papery molted skeletons of sand crabs that often litter the beach along the wrack line. As arthropods grow, each must periodically cast off its old exoskeleton and develop a new, larger one, hence the molts. The little sand crabs responsible for them are burrowed in the sand.

These animals do not feed with pincers like most other crabs but rather extend their long feathery second antennae above the sand during a wave's backwash to trap minute water-borne food particles. These antennae lie lateral to the eyes and when not in use are folded out of sight, as they are in our photo. The first antennae, between the eyes, are used as a snorkel, allowing the crab to breathe when it is buried in the sand (Jensen 1995). During incoming tides, sand crabs allow themselves to be carried up in the swash of a breaking wave; then, as the wave recedes, they burrow tail-first into the sand and extend their second antennae to feed. Moving up and down the beach with the tides extends the crab's feeding time to virtually subtidal continuity.

Emerita's streamlined body provides minimal resistance to movement through water and sand. Sand crabs have lost the ability to walk, but they use their legs to dig backward into slurried sand, shoveling sand rapidly forward from underneath the body (Faulkes and Paul 1997). In addition to burrowing, the animals also swim in reverse by sculling with their rear legs.

Notice the stalked compound eyes, the small first antennae, the strong, stumpy, oarlike legs (visible in the ventral view of the female), and the lack of food-tearing "crab" pincers. These crabs and their relatives (anomurans, a group that also includes hermit crabs) at first appear to have only four legs on each side. There are fifth legs, however; much reduced, they are tucked inconspicuously under the chitinous shell and serve as gill cleaners.

Sand crabs are preyed upon by shorebirds during the waves' backwash and by surfperches and some crabs when the lower beach is submerged. In Carmel Bay we often have seen a young sea otter diving in the breaking waves; this predator, too, may well be taking these little crustaceans.

Female *Emerita* bearing eggs, about 30 mm long. Carmel Beach.

The underside of this female shows orange eggs that she is brooding between her thorax and her folded abdomen; her abdominal swimmerets hold the eggs in place.

Males, about half the size of females, reach sexual maturity soon after they settle and metamorphose out of their larval stage. Most then perish in their first winter (Efford 1967). In contrast, females overwinter in large numbers and reproduce again in their second year, so the ratio of males to females varies with the shift of the seasons.

In early summer, when a female is fertile, two or three males attach themselves to her. The breeders are united firmly enough to stay together as they are carried back and forth by the waves. The males then deposit ribbons of sperm beneath the female's body, and fertilization occurs as she extrudes her eggs several hours later (MacGinitie and MacGinitie 1968). The mother carries her eggs for weeks or even months before they hatch and the larval young swim away. The larvae molt several times before settling into a benthic life, and the juveniles continue to molt as they grow, supplying the wrack line with fresh skeletons.

Emerita analoga occurs not only on California's beaches, but also on those of Chile at a similar latitude in the southern hemisphere. They are not found in the intervening tropical areas. There is no fossil record of the species, but genetic studies suggest that both the northern and southern populations probably arose from a common ancestor and then, most likely as drifting larvae, dispersed widely, though certainly under geologic and climatic conditions very different from those that exist at present. Today enough genetic differences have accumulated to regard them as separate species even though they still bear the same name (Tam, Kornfield, and Ojeda 1996).

Thinopinus pictus
Phylum Arthropoda / Class Insecta

Rove beetle, 18 mm long. Carmel Beach.

Cafius seminitens
Phylum Arthropoda / Class Insecta

Above right: Rove beetles, 10 mm long. On decomposing blade of red alga, Carmel Beach.

These beetles are among several species of rove beetles that live on sandy beaches where piles of kelp collect at the high tide level. You won't see these flightless beetles unless you explore the damp sand amid the decomposing kelp at night (with the aid of a flashlight) when the tide is low. During the day and at high tides *Thinopinus* burrows deeply into the sand (its legs are adapted to digging as well as catching prey), and *Cafius* hides under piles of damp kelp. In the beam of your light, the relatively large *Thinopinus* beetles blend with the sand when they are immobile, but they really scamper when prodded. Carnivores, they quickly lunge forward as they use their formidable mandibles to grasp their prey—beach hoppers, sand isopods, flies and their larvae on the decomposing kelp, and the fingers of unwary naturalists. Both adult *Cafius* beetles and their larvae feed voraciously on the maggots and pupae of kelp flies, on amphipods, on various other scavengers, and on other *Cafius* species (Evans 1980).

Neither species can swim, and they must breathe air. Nevertheless, they can survive brief submersion in seawater by becoming quiescent and reducing their oxygen consumption. When reexposed to air they make up their oxygen debt. In one heartless but dramatic experiment, half of an experimental group of *Thinopinus* beetles succumbed when submerged for thirteen hours in seawater, and it took the lucky survivors 50 minutes to recover from the ordeal (Topp and Ring 1988). *Cafius* is more tolerant of submergence. In any event, by sticking to the beach at the high tide level these rove beetles avoid such extreme stresses.

Lottia digitalis / austrodigitalis
(= Collisella digitalis)
Phylum Mollusca / Class Gastropoda

Ribbed limpet, 25 mm long. High intertidal zone, vertical wall, Carmel.

In the high rocky intertidal splash zone well above the range of predatory snails and seastars we find these large ribbed limpets, typically aggregated on vertical surfaces, as these are. Their orientation on these vertical rocks provides some protection not only from heat and desiccation, but also from predation by the black oystercatcher, *Haematopus bachmani*, which pries limpets from rocks with its long chisel-shaped bill (Frank 1982). We have noticed that when this bird feeds at the water's edge, where limpets may be easier to pry off, it invariably slithers about on the wet slope, unable to get a foothold. In less awkward spots, it cannot reach higher than about 35 cm above its perch (Hahn and Denny 1989). The limpets in the photograph were 1.5 m above a ledge, well out of an oystercatcher's reach.

At high tides and during rough water, limpets disperse on the soaked rocks and graze on microscopic algae, particularly diatoms. They favor diatoms that grow in chains, apparently because the limpet's radula (its scraping tongue) can detach one end of the chain and pull it in like spaghetti, whereas unlinked diatoms hang on more tenaciously (Nicotri 1977). The diatoms' glassy enclosures appear in the limpets' feces. When they are immersed or wet with spray, limpets breathe through a feathery gill that lies in a shell-protected cavity over the head. While exposed at low tides, they suffuse their mantle (the thick membrane that encloses the internal organs) with blood and use moist body surfaces to absorb oxygen and release carbon dioxide. *Lottia* tends to return to its home base at low tide but not always to exactly the same spot as some limpets do. All limpets broadcast both sperm and eggs to the sea, where the eggs are fertilized and a new generation begins to develop into tiny larvae.

This species closely overlaps in range with the rough limpet, *Macclintokia scabra*, but the latter tends to venture out onto horizontal surfaces, oystercatchers be damned.

Macclintokia scabra
(= Collisella scabra)
Phylum Mollusca / Class Gastropoda

Rough limpet, 20 mm long. High intertidal rock, Carmel Bay.

These limpets are on a very exposed, horizontal surface; note how well they blend in with the granite rock. Though camouflage may provide them some protection, more importantly they fit their home scar so tightly that they are difficult to dislodge. This is an environment of temperature extremes, but the rough limpet possesses more "heat-shock" proteins (that repair damage to thermally injured proteins) than do other species that live in more sheltered sites (Sanders et al. 1991), and so can better tolerate high temperatures.

Like the ribbed limpet, *Lottia digitalis/austrodigitalis*, which is often a close neighbor, these limpets disperse at high tides to graze on microscopic algae. They are active when the rocks are washed by incoming and outgoing tides. At low tide, most of them return by exactly the same route to exactly the same spot and orientation. Thus, as this photo shows very well, the corrugated rims of their shells become sculpted to precisely fit irregularities in the granite at their home base. On softer rocks, the rough limpets excavate a depression with their radula, aided by acidic secretions, to secure a perfect fit.

The ability of the rough limpet to hang on has been measured with a device that strikes a sharp horizontal blow, simulating the peck of the black oystercatcher. And as one might expect of a species that lives at such exposed sites, it is better able than other limpets (such as the ribbed limpet) to withstand such a shearing strike (Hahn and Denny 1989). The tenacity of limpets involves suction within a tight seal that they make on a smooth surface; if they are forced to glide over a small hole so that the suction is broken, they can more readily be detached (A. Smith 1991). So camouflage, preparation of a secure home site, and physical tenacity all may help explain the oystercatcher's relative lack of success in preying on rough limpets.

Limpets, like many other snails, devote considerable energy to producing the mucus needed to adhere to the rock as they move about. The rough limpet's mucus trails and those of some other limpets (e.g., *Lottia gigantea*) nourish microalgae. By retracing its wanderings, this limpet can graze on the diatoms and other algae whose growth its trails have stimulated, thereby recouping some of its metabolic expenditure (Connor and Quinn 1984).

Both the rough and the ribbed limpets compete for algal food. If the latter species is excluded from their overlapping ranges, the rough limpet will grow larger than it otherwise would (Haven 1973).

Haematopus bachmani
Phylum Chordata / Subphylum Vertebrata / Class Aves

Black oystercatcher, 20 cm high. Intertidal rocks, Monterey Bay. (Photo courtesy of Ronald Branson)

Just as fish do at high tides, many birds scour the intertidal shore of Monterey Bay at low tide—several gulls, a dozen or so common shorebirds, even herons. The black oystercatcher, a nonmigratory "rocky shorebird," is black and big as a crow, with a chisel-like vermilion red bill, red-rimmed yellow eyes, and pink legs—altogether a bizarre sight. It breeds just above the high intertidal zone, usually on offshore rocks where the nests are not preyed upon by rats. It forages at low tides, principally on mussels and on limpets such as the ribbed limpet (*Lottia digitalis/austrodigitalis*), rarely on the more tightly adhering rough limpet (*Macclintokia scabra*). It also takes shore crabs, chitons, isopods, and small unshelled invertebrates. In winter it may forage in small flocks. At any time of year its piercing call can often be heard above the noise of the pounding surf.

Mussels are vulnerable to oystercatcher predation as they gape open when splashed by a rising tide and again when they are exposed by the ebbing tide. During its attack on mussels, the bird delivers a sharp stab between the valves, rapidly levers them open, and withdraws the meat. It apparently has to sever the shell-closing adductor muscle with the initial stab; otherwise, the bird must struggle to withdraw its bill from between the valves. Should the mussel come loose, the bird carries it to higher ground to eat it.

Limpets in shallow tidepools are vulnerable when they are not deeper than bill length. Sharp jabs at the edge of the shell dislodge the limpet. The bird then places the shell upside down, bites around its edge, picks up the body of the limpet and shakes the shell free, and swallows the meat in one gulp.

Mastocarpus papillatus (= *Gigartina papillata*)

Phylum Rhodophyta / Class Florideophyceae

Left: Foliose red alga, about 10 cm long. Mid-intertidal rock, Carmel Bay.

Scattered on intertidal rocks, *Mastocarpus papillatus* occurs as curly, stiff, dark red to black leaflike blades. This one dangles in a tidepool at high tide. The blades appear annually, but intense summer grazing by snails and then the battering of winter storms prevent them from surviving year round. They are vulnerable to grazing in part because a blade can be detached easily at the holdfast. These algae have a high photosynthetic rate and grow to replace themselves rapidly, but eventually the herbivores make inroads.

Mastocarpus has a strange heteromorphic life cycle, in this case an alternation between gamete-producing blades and spore-producing dark crustose "tar spots" (Fig. 4). (For a discussion of algal life cycles, see the introduction.) Female gametophytic blades, depicted here, are bumpy with little papillae, whereas male gametophytes have smooth blades. Female gametophytes produce eggs in the papillae but do not release them. These eggs are fertilized in place by passively drifting sperm released by male gametophytes. Here, on the papillae of the maternal blade, the diploid third phase in a red alga's life cycle (the carposporophyte) develops. This third phase in turn releases diploid carpospores, which settle and grow into tarlike sporophytes.

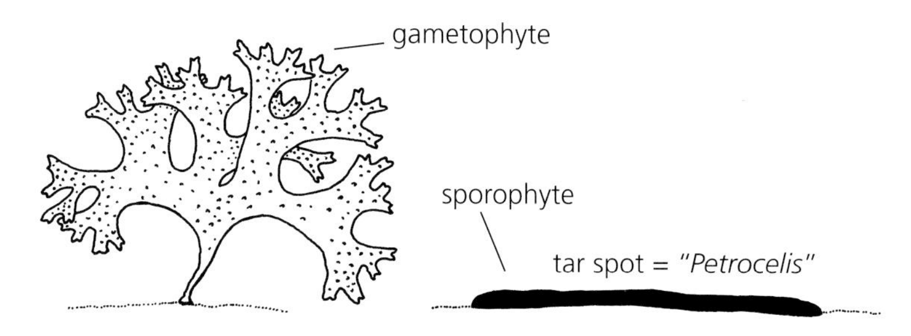

Fig. 4. *Mastocarpus papillatus* life cycle

Mastocarpus papillatus (crust) (= *Petrocelis middendorffii*), about 12 cm. Mid-intertidal rocks, Carmel Bay.

This dark, crustose tar-spot alga is, despite the fact that it often still goes by a separate name, actually the sporophyte phase of *Mastocarpus*'s life cycle. It wasn't until 1975 that spores from "*Petrocelis*" tar spots were cultured; they were found to grow into *Mastocarpus* gametophyte blades, thereby establishing the rather implausible relationship between these utterly disparate forms (Polanshek and West 1977).

The tar spots of *Mastocarpus* form on mid-intertidal rock, grow slowly, and may live for many decades. Meiosis (reduction division) yields haploid spores. When the spores settle, they divide and grow to become the male or female gametophytic blades we just looked at.

Why this heteromorphic life cycle? It appears that the alga is "hedging its bets." Should a few years of adverse conditions, such as heavy grazing by limpets, wipe out the foliose, annual gametophytes, the species can reproduce from the long-lived crustose sporophytes. *Mastocarpus papillatus (crust)* appears to be quite resistant to grazing, sand scouring, death by burial, and wave shock (Lobban and Harrison 1994). In fact, judging by its photosynthetic rate, which is only one-fifth that of the gametophytic blades, it seems to put more energy into structural and chemical defenses than into its photosynthetic machinery (Zupan and West 1990). Moreover, *M. papillatus (crust)* apparently survives better, even thrives, in the presence of grazers, such as the little limpet in the photo, which may keep other algae from overgrowing and shading it (Slocum 1980). This tar spot certainly has a very clean surface.

In central California, adding complexity to the already complex, this alga may completely bypass the crustose sporophyte phase; many female gametophytes form carpospores that then develop directly into more female gametophytes (Zupan and West 1988). One advantage of this shortcut is evident: the life cycle then requires only a single patch of open space and a single "establishment event" to persevere to the next generation. But they don't all do it, for there are still plenty of tar spots about Monterey Bay.

Fucus gardneri (= F. distichus)
Phylum Heterokontophyta / Class Phaeophyceae

Rockweeds, about 12 cm. Intertidal rocks, Carmel Bay.

Fucus is a perennial brown alga that may dominate mid-intertidal rocks. In this photo it is surrounded by a red alga and another species of rockweed. At low tide the exposed rockweeds partially dry out—and yet, remarkably, that is when their rate of photosynthesis is greatest (Johnson et al. 1974). We have seen that another intertidal alga, *Mastocarpus,* has a particularly tough, herbivore-resistant phase in its life cycle, *Mastocarpus papillatus (crust). Fucus,* in contrast, deters predation by chemical means: its "juice" is high in toxic compounds (polyphenolics), which snails and urchins avoid (Lobban and Harrison 1994).

Rockweed reproduction is a fascinating subject. Like the kelp *Cystoseira,* a subtidal alga we discuss later, rockweeds are diploid. When they are reproductive, the blade tips become swollen with conceptacles—cavities containing reproductive cells—visible in the photo as little bumps. Within these conceptacles meiosis occurs, to produce ova or sperm. The eggs and sperm are released from different conceptacles on the same algal body; subsequent fertilization and germination then leads directly to new rockweeds. Because the rockweed *Fucus* has no separate sporophytes or gametophytes, its life cycle resembles those of animals, at least genetically.

Like so many encounters in the marine world, fertilization in *Fucus* is not a matter of chance alone. As early as 1854, M. G. Thuret observed under a microscope the influence of eggs on the sperm in a species of *Fucus* (quoted by Maier and Müller 1986): "Il est difficile, quand on observe ces phénomènes avec attention, de ne pas se laisser aller à croire qu'une impulsion particulière dirige les anthérozoïdes vers les corps qu'ils doivent féconder" (It is difficult, when one observes these events carefully, not to believe that some specialized force of attraction directs sperm toward the eggs they will fertilize). It is only in recent years that sensitive analytical methods have identified a number of pheromones in brown algae that explain Thuret's remarkable observation. These volatile compounds are secreted by the rockweed eggs in extremely dilute concentrations (10^{-12} to 10^{-14} mol), starting shortly after the eggs are released and ending with fertilization. It appears that these chemical signals may attract sperm from several millimeters away (Maier and Müller 1986). Thuret no doubt would be delighted to know that his interpretation of these events was confirmed by scientists almost a century and a half later.

But even more sophisticated mechanisms are at work in these seaweeds as they set about reproducing. Studies of *Fucus* species show that peaks of egg and sperm release occur only during the day, at low tides, when the ocean is calm. Under such conditions, fertilization success is an astonishing 78–100 percent (Serrão et al. 1996)! It is as if the algae "want" to release their eggs and sperm when photosynthesis can occur, when sperm concentration will not be diluted by rough water, and when the gametes are concentrated as they float on the surface.

Following the *Fucus* egg's fertilization and the creation of an embryo, the first cell division occurs asymmetrically, producing daughter cells with different fates: one ultimately produces leaflike blades, and the other, a rootlike holdfast. But these two cell lines, blade and holdfast, can change their identities. If a cell destined to develop into a blade touches an empty holdfast cell wall, it will change into a holdfast cell. Far from being an inert box, then, the cell wall apparently produces a signal that determines the ultimate fate of cells during development (Bouget, Berger, and Brownlee 1998). Within four hours of fertilization, the holdfast cell of the two-celled embryo secretes a specialized "glue" that is transported to the top of the holdfast tendril, enabling the embryo to attach to the substratum (Vreeland, West, and Epstein 1995).

Fucus embryos have become a well-studied experimental system. These studies should cause us to look at this rockweed with newfound respect for its sophistication.

Pollicipes polymerus
Phylum Arthropoda / Class Crustacea

Leaf barnacles, about 7 cm, with California mussels, *Mytilus californianus.* Mid-intertidal zone, low tide, Carmel Bay.

On surf-swept intertidal rocks of the open coast, stalked leaf barnacles often live tightly packed together with California mussels, as shown here. The stalk of this barnacle contains the ovaries, and at its base (actually the top of the head!) is an adhesive gland that glues the barnacle to its substratum. The rest of the internal organs and the appendages are located above the stalk in a cavity surrounded by an elaborate array of armored plates that close tightly at low tides. In contrast to acorn barnacles, which feed by rhythmically sweeping the water with their feathery legs, leaf barnacles extend strong, hairy legs steadily into the backwash of waves, trapping crustaceans up to 10 mm in size and large particles of detritus.

Gulls are major predators of leaf barnacles when the tide is out (Meese 1993); probably fish and certainly the ochre seastar *Pisaster ochraceus* prey on them when the tide is in. By reducing the population of *Pollicipes*, gulls indirectly foster an augmented population of the coexisting and space-competitive California mussel (Wootton 1992, 1994). Both barnacles and mussels are suspension feeders, although barnacles trap much larger objects with their legs than mussels select with their gills.

Settled barnacles would seem to epitomize the notion of being stuck in place, and this is certainly true of acorn barnacles like *Balanus.* Tiny *Pollicipes*, however, attach to the stalks of adults when they first settle, then creep very slowly toward the substratum (Woll 1997). Once there, they maneuver themselves into a permanent and efficient orientation to exploit the wave-swash for food.

Although mussels and barnacles live together and both are suspension feeders, they differ radically in their reproductive habits. Individual mussels are either male or female and broadcast their gametes to the sea for external fertilization. Barnacles, in contrast, are hermaphroditic and mate by copulation. Apparently not self-fertile, they must aggregate so that they can cross-fertilize with nearby individuals. Their penises are often remarkably long–five or six times their shell size—and sometimes are mistaken for worms meandering among the massed bodies! One *Pollicipes* barnacle may produce up to seven broods per year, with up to 240,000 larvae in one brood. Feeding larvae (nauplii) are shed to the sea and molt several times before metamorphosing into non-feeding larvae (cyprids) whose one goal in life is to settle, metamorphose, and cement themselves head down next to their own kind. Full-grown *Pollicipes* barnacles may be twenty years old (Newman and Abbott 1980).

Mytilus californianus
Phylum Mollusca / Class Bivalvia

Below: California mussel, about 9 cm. Intertidal rocks, Monterey Bay.

Many surf-swept intertidal rocks in both Carmel and Monterey Bays are covered with extensive beds of mussels. The mussels in this photograph are surrounded by the green alga *Enteromorpha.* As we noted at the beginning of this chapter, the upward extension of intertidal organisms is limited principally by physical factors and their extension deeper into the sea is limited principally by predation. Mussel predators include man, sea otters, ochre seastars, black oystercatchers, shore crabs, and predatory snails. Sea otters, avid hunters of invertebrates, do most of their foraging subtidally, but occasionally they seek food in low intertidal waters, sometimes with profound effects on mussel communities. Off Hopkins Marine Station in Pacific Grove, for example, there once existed a population of very large mussels, many in the low intertidal zone—hence the site's informal name of Mussel Point—but no longer. In about 1963, following the return of the sea otter to Monterey Bay, the lowest large mussels were completely cleaned out in a very few months (Judson Vandevere, pers. comm.). Of course poachers, who made off with the rest, didn't help either! The ochre seastar is another predator that may prevent mussels from extending their range much below the lowest tides. In addition, surf and the ochre seastar often dislodge mussels, which then become a major food item for the giant green sea anemones (*Anthopleura xanthogrammica*) that lie in wait in nearby surge channels.

Mussels attach themselves to the rock by strong filaments (byssal threads) secreted by a gland at the base of the foot. Unlike other molluscs, mussels and their bivalve relatives do not have a rasping, tongue-like radula and jaws. Instead, with its valves slightly agape, a mussel pumps seawater into the mantle cavity, which it then filters through its gills. This water flow is generated by the beat of millions of cilia on these gills. The mucus-laden gills absorb oxygen and sort and trap food particles such as detritus and phytoplankton, which then are passed along elaborate grooves to the mouth.

Like many bottom-dwelling invertebrates, mussels synchronize their spawning with phytoplankton blooms. This ensures that the mussels' larvae will develop when there is plenty of microscopic food for them to eat. It also minimizes larval mortality, as predatory zooplankton are at a yearly low at the onset of the spring phytoplankton blooms (J. Smith and Strehlow 1983; Starr, Himmelman, and Therriault 1990).

In California, all bivalve molluscs are quarantined from May through October because of "red tides," blooms of dinoflagellates and diatoms that are especially likely to occur during these months. Several species of dinoflagellates produce a potent neurotoxin that shellfish like California mussels assimilate and temporarily store. When people eat these contaminated animals, paralytic shellfish poisoning results. Initially the toxin causes numbness and ting-

ling about the mouth, but symptoms can progress in severity and even cause death. Although neither sea otters nor black oystercatchers are immune to the toxin, somehow these species learn to avoid contaminated shellfish (Kvitek, DeGange, and Beitler 1991; Kvitek, Bretz, and Thomas 1999).

In September 1991 the neurotoxin domoic acid, produced by a species of diatom, caused a notable die-off of pelicans and Brandt's cormorants in Monterey Bay (Garrison et al. 1992). The birds had fed on anchovies that had acquired the toxin directly from the diatoms. And in 1998, along the central California coast, over four hundred California sea lions died and many others showed neurological disturbances; they too had fed on contaminated anchovies. Other die-offs of marine mammals exhibiting similar behaviors have been documented in recent decades (Scholin, Gulland, Doucette et al. 2000). Humans, not surprisingly, are also vulnerable to such poisoning. In late 1987 in eastern Canada, after eating mussels contaminated with domoic acid, a number of people became ill with systemic and neurological manifestations, and four of the most seriously ill died. Some of the survivors had predominantly short-term memory impairment, which prompted the nickname "amnesic shellfish poisoning." The relative preservation of other cognitive functions helped to distinguish them from patients with Alzheimer disease (Teitelbaum et al. 1990; Clark et al. 1999). Monterey Bay may yet be a hot spot for domoic acid–producing blooms, but now a program is in place to warn us.

The U.S. Mussel Watch Program uses *Mytilus californianus* and several other species of bivalves to detect minute but potentially dangerous levels of environmental toxins in seawater, among them organic pollutants and heavy metals such as lead and cadmium. The mussels are to nearshore waters what canaries once were to coal mines: sensitive monitors of chemical troubles in the environment. A moderate-sized mussel bed can filter an estimated 24 million metric tons of seawater in a year, concentrating pollutants in the animals' tissues all the while. In 1984 Mussel Watch detected a high level of lead in Monterey's inner harbor, the product of leakage from a lead slag heap—since removed (Flegal, Rosman, and Stephenson 1987). The one live mussel that could be found in the inner harbor then had a lead concentration of 1,826 parts per million; normally there should be less than 0.5 part per million.

This monitoring program has limitations, however. Mussels have defenses that protect them from many potentially harmful environmental substances, enabling them to expel such substances or detoxify them. Moderately soluble organic compounds, for example, such as the pesticide PCP (pentachlorophenol), bind to a special mediating compound known as P-glycoprotein and then are transported out of the mussel tissue (Cornwall et al. 1995). In contrast, DDT and PCB (polychlorinated biphenyl), *completely insoluble* compounds, are not affected by the transport mechanism and readily accumulate within the animal. Thus, monitoring for environmental pollution with *Mytilus* might fail to reveal the presence of PCP, though it would readily detect DDT and PCB.

Pisaster ochraceus
Phylum Echinodermata / Class Asteroidea

Ochre seastars, field of view about 50 cm. At high tide, Carmel Bay.

Provided it can stay more wet than dry, the ochre seastar is admirably adapted to rugged intertidal habitats that expose it to severe stresses. It hangs on so tenaciously that many of its tube feet will tear off before the animal can be pulled from a rock. Despite its name, many of these seastars are purple rather than yellowish. Here we see *Pisaster*, immersed at high tide, together with stalked and acorn barnacles, California mussels, and foliose and coralline red algae. Ochre seastars prey heavily (whenever the tide is in) on barnacles and mussels, and they, together with sea otters and fishes, may limit mussel beds from extending deeper—which is just what *Pisaster* appears to be trying to do here.

Ochre seastars, field of view about 50 cm. Intertidal zone, Carmel Bay.

How do ochre seastars get inside the seemingly well-defended mussels they so often attack? After all, most people find it a struggle to open a mussel's valves even with a knife. *Pisaster*, and indeed most seastars, feed by grasping their prey, turning part of their stomach inside out, extending it from the mouth, and inserting it into their victim to digest the prey's tissue. That's well and good for dining on snails and barnacles, but getting one's stomach inside a mussel presents a special challenge. *Pisaster* may insert some of its thin, everted stomach, or at least introduce its digestive enzymes, through a narrow slit where the mussel's anchoring byssal threads pass between its valves. And it can pull the valves slightly agape with powerful, sustained traction of as much as 4000 grams, exerted through its tube feet while it stiffens its arms. Finally, it can exploit the mussel's need to open its valves periodically to breathe or feed, and more readily insert its stomach between the valves then. Once its stomach is inside a mussel, of course, it's a piece of cake. One ochre seastar can consume up to eighty mussels per year (Feder 1955).

Predators on *Pisaster* include gulls and sea otters. But, like all animals, it has microscopic predators as well, most notably infectious agents such as microbes. In 1882 a young Russian zoologist, Elie Metchnikoff, pierced a tiny, transparent seastar larva with a rose thorn. The next day he found that innumerable amoeba-like defensive cells were attempting to engulf the thorn. For his insight into the meaning of this event and for his subsequent studies, Metchnikoff is recognized today as the father of cellular immunology. Now, over a hundred years later, we have learned a great deal about the reaction he investigated. The seastar's defensive cells produce a chemical signal that causes them to multiply, to swarm to the site of any microbes introduced in a wound, and to engulf the invaders (Beck and Habicht 1996).

A very similar reaction occurs in vertebrates. Should a gardener stick a finger with a rose thorn, her immune system will promptly release a chemical signal, interleukin-1, that mobilizes defensive cells to overcome the invading microbes, a response much like that which Metchnikoff observed in seastars. The chemical signal from *Pisaster*'s cells, moreover, can be blocked with antibodies to mammalian interleukin-1, evidence that these molecules are very closely related (Burke and Watkins 1991).

The echinoderms and chordates (including humans) share the same early embryonic developmental pattern in which the mouth forms secondarily (deuterostome development); the two groups probably diverged from each other some 600 million years ago in late Precambrian times (Ayala et al. 1998). And now we learn that the same defensive molecule, interleukin-1, has been conserved in these two groups all these years, testimony to its fundamental physiological importance.

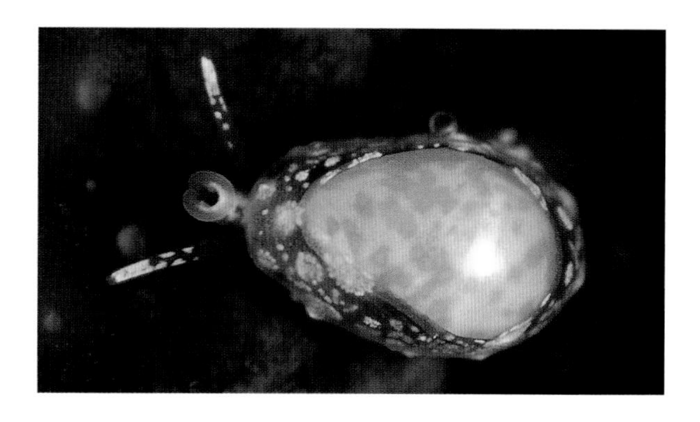

Granulina margaritula
Phylum Mollusca / Class Gastropoda

Margin snail, 2 mm long. From tidepool, Carmel Bay.

It takes a patient search in tidepools to find this tiny snail, hidden among green algae, as it is here, or among grains of sand. But it's worth the effort. Look for a mottled, creamy, shiny shell; you'll want a hand lens to be sure you have a snail and not a grain of sand. The genus name *Granulina* is a Latin word meaning "little pearl"—a good choice. Margin snails have internal fertilization, and feeding larvae result; however, little else is known of *Granulina*'s biology. Here you see just a hint of its beautiful mantle, the protective membrane that expands to cover its shell.

Leptasterias pusilla
Phylum Echinodermata / Class Asteroidea

Six-armed seastar, about 20 mm across. From tidepool, Carmel Bay.

This little seastar, no bigger than a nickel, lives in tidepools in the mid and low intertidal zone, almost invariably hidden under rocks during daylight hours. It forages at night, taking small snails and, in the laboratory at least, small chitons, mussels, and barnacles. *Leptasterias* is vulnerable to desiccation and, given that its color tends to match the substratum, probably to visual predators. In the field, a gull was seen to take *Leptasterias* from a tidepool, and in an aquarium an octopus turned over a brooding *Leptasterias* and stole her eggs (R. Smith 1971). The little seastar tolerates tidepools that are diluted by rain or warmed by sunlight.

In January females produce forty to one hundred yolky eggs that are fertilized by spawning males. The females hunch up over their egg mass, holding them in place beneath the mouth with their tube feet while turning them and ventilating them with water. The developing embryos require this attention and do not survive removal from their mother. Development is direct, and by April and May tiny young crawl away from their mother, looking like tiny, moving snowflakes. With their departure, the mother again is able to feed.

Below: Granulina pursued by *Leptasterias.* From tidepool, Carmel Bay.

We have brought predator and prey together in our lab, and the action, as seen under a dissecting microscope, is explosive. Here, the little margin snail instantly senses *Leptasterias* chemically and glides away from it with incredible rapidity, raising its mantle at the same time. The photograph cannot show the speed of the retreat, but it does show the colorful beauty of this minute snail's protective mantle. A snail's mantle not only presents a slippery surface that may foil a seastar trying to get a grip, but it may secrete noxious chemicals as well. *Leptasterias* shows no sign that it detects *Granulina* at a distance.

Aplysia californica
Phylum Mollusca / Class Gastropoda

Sea hare, about 15 cm. Monterey Bay, 10 m deep.

Aplysia, unlike most sea slugs, is an herbivore that forages on red algae and on sea grasses as it roams about between intertidal and subtidal habitats. Perhaps it does suggest a hare with its "rabbit ears" extending up in a V. These snails are short-lived but rapidly growing; some may attain a length of 40 cm and a weight of several kilograms during their life span of only about a year. Looking like fleshy, slimy blimps, they lack an external protective shell. In southern California, lobsters and the predatory sea slug *Aglaja* prey on small *Aplysia* (Pennings 1990a), as does the giant green anemone, although toxic portions are regurgitated (Winkler and Tilton 1962). Large sea hares seemingly have no predators, however.

This *Aplysia* explores the sand amidst an eelgrass bed near Del Monte Beach. As it advances, it slowly waves its head from side to side, receiving sensory input through its oral tentacles, its rhinophores (the "rabbit ears"), and its eyes; in addition, the whole body is sensitive to touch. Here, one oral tentacle lightly explores the sand; both the oral tentacles and the rhinophores will detect a nearby red alga's amino acids at a distance. The tiny eyes tell *Aplysia* that it is daylight, time to be active. The eye consists of a cornea, a lens that is partially surrounded by the retina, sensory neurons, and an optic nerve that leads to a cerebral ganglion. The succession of day and night entrains a persistent circadian rhythm within the eyes themselves, as manifested by electrical activity in their neurons. When removed from the snail's body, the eyes retain the circadian rhythm for several days, even in the dark. In the laboratory, the circadian clock can be reset by altering the cycle of light and dark, both in the intact animal and in isolated eyes (Kandel 1979).

Aplysia, not surprisingly, is a very, very well studied animal. Its central nervous system consists of some twenty thousand neurons clustered in five interconnected major ganglia. The large size of the neurons and their distinctive colors lend themselves to ready identification and to detailed functional studies. As many as one hundred biologically active peptides may mediate neuroendocrine function. For example, one family of genes, active in well-identified ganglia, encodes peptides that mediate a whole series of behaviors during egg laying: the snail's locomotion ceases, feeding is inhibited, respiratory pumping through the mantle cavity increases, and head waving begins, which facilitates winding of the egg string as it exits the genital pore (Scheller et al. 1984). Imagine getting a handle on this complex series of events!

Mating aggregation of *Aplysia*. Southern California, 15 m deep.

These hermaphroditic snails are solitary most of the year; at certain times, however, particularly in the summer and autumn, they gather in large aggregations of mating and egg-laying individuals. In the scene from which this detail comes, some six or eight sea hares are copulating, each with its penis in the vaginal orifice of the one ahead. Such aggregations may last for several days, but there is a constant flow of individuals in and out. New animals may join the group by acting as a sperm donor to the terminal male. Many of the egg-laying animals simultaneously mate as sperm recipients (Pennings 1991). Eggs are laid in enormous numbers, piled up in long yellow gelatinous strings. MacGinitie (1934) calculated that one large animal laid almost five hundred million eggs during its few reproductive months. The eggs release a diffusible peptide pheromone "attractin" that lures other *Aplysia* to the party (Painter et al. 1998). The larvae are planktonic for a month before settling and metamorphosing—often, but not necessarily, on the red alga *Plocamium*.

Red alga *Plocamium cartilagineum*, field of view 15 mm. Carmel Bay, 15 m deep.

Plocamium, a small, delicately branched red alga, plays an essential role in the life of *Aplysia:* the newly metamorphosed snails require this alga to support growth, though they can *survive* for a time on a number of other algae. On Santa Catalina Island, Steven Pennings found small *Aplysia* measuring 10–30 mm—the majority and the most vulnerable of the population—almost exclusively on *Plocamium.* Their small, relatively weak mouthparts probably limit grazing on coarser algae, though they readily take chopped up pieces of larger, tougher algae. Older, larger, and less vulnerable *Aplysia* gradually broaden their diet to include many green and red algal species as well as seagrasses; they even can be reared on romaine lettuce. For *Aplysia,* the delicate *Plocamium* has benefits that go beyond supporting growth. By eating *Plocamium,* as opposed to almost any other alga, the snail acquires chemicals (brominated terpenes) that accumulate in the skin and internal organs and provide a passive defense against predators. For example, when fed to kelp bass, a predatory fish, *Aplysia* raised on sea lettuce, *Ulva* (which lacks terpenes), is more vulnerable than *Aplysia* raised on *Plocamium,* which causes these fish to regurgitate (Pennings 1990a, 1990b).

Aplysia, about 9 cm, inking. Tidepool, Carmel Bay.

One day while exploring tidepools at low tide we found this partly emersed sea hare ejecting swirling clouds of ink from its siphon, apparently spontaneously—at least, we could find no evidence of molestation. Many other sea hares of comparable size were busily grazing on red algae and surfgrass in adjacent pools. All these animals had outgrown their vulnerable youth; however, for small *Aplysia* (less than 30 mm in length) this purple ink plays a very important role. Exuded when the snail is under attack, the ink is probably even more effective than the passive defense of terpene-laden flesh. *Aplysia* fed on red algae secrete copious amounts of the ink when ensnared by a giant green anemone's tentacles and ingested, causing the anemone to eject both the ink and the sea hare from its digestive cavity. Sea hares that have been reared on the green alga *Ulva* are inkless and much more likely to succumb under similar circumstances. And snails with only passive defenses—terpene-laden flesh—are more vulnerable than those with only an active defense—ink but no terpenes (Nolen et al. 1995). *Aplysia*'s ink also acts as a warning pheromone: by avoiding it, the snails distance themselves from a site of potential danger.

Aplysia obtains its ink from the accessory photosynthetic pigment phycoerythrin, the pigment that makes red algae red. Having freed phycoerythrin from algal cells in the foregut, *Aplysia* digests the pigment in specialized stomach cells, thereby producing ink. After storage in an ink gland, this remarkable substance is available for its defensive role (Coelho, Prince, and Nolen 1998).

Ampithoe sp.

Phylum Arthropoda / Class Crustacea

Below: Amphipod, about 12 mm long, on sea lettuce, *Ulva* sp. Low intertidal rocks, Carmel Bay.

We found this beautifully camouflaged amphipod among blades of *Ulva* in the low intertidal zone. Because the cuticle covering an amphipod's body is translucent, the animal's color derives from a variety of factors, including a pigment deposited in the cuticle, chromatophores (pigment cells that can be altered to produce color change), the color of the blood and internal tissues, and even food in the gut. *Ampithoe*'s red eyes glare out against its otherwise almost perfect color camouflage.

The eyes' bright red color is due to rhodopsin, a visual pigment that absorbs a spectrum of light's electromagnetic energy and transforms that energy into a neural stimulus. The pigment is present in all known animal photoreceptors—*Ampithoe* couldn't see without it. In this case, the photopigment appar-

ently absorbed blue-green light; it thus reflects all that is left over, predominantly red.

The surface of an arthropod's compound eyes has many tiny hexagonal facets, barely visible in this photo. Each facet is the surface of a thick lens at the broad end of one long, conical optical unit, or ommatidium. At its inner, apical end, each ommatidium has several light-sensitive cells that, together with the visual pigment, form a single photoreceptor unit. Each photoreceptor unit innervates neurons that lead to the animal's optic nerve and to the brain (Fig. 5).

Ommatidia have slightly different but overlapping visual fields, and a moving object successively stimulates one ommatidium after another. A crustacean's compound eyes are particularly good at detecting the slightest motion, but the design is ineffective at resolving detail, and there is no means to change focus. Shore crabs, for example, can instantly detect any object's approach and will respond by disappearing into a crack, but probably they have only a rough perception of the approaching object's finer features. One ingenious experimenter succeeded in photographing Darwin's portrait through a firefly's compound eye (Land 1980); the resulting image is barely recognizable as a human face—though perhaps the insect's brain is able to fine-tune its ommatidial data in ways Land's camera could not.

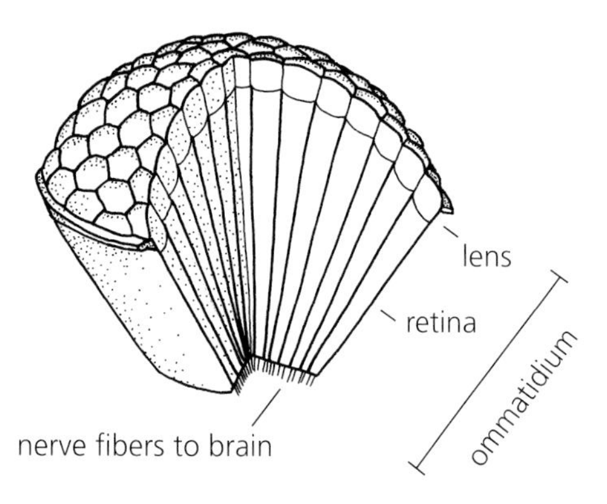

Fig. 5. Arthropod compound eye (after R. Hesse, in Pearse et al. 1987)

Phyllospadix scouleri

Phylum Spermatophyta / Class Angiospermae

Opposite: Scouler's surfgrass and brown and red algae, field of view about 1.5 m. Low tide on the open coast.

It's a quiet day and very low tide on the open coast. Glistening surfgrass (*Phyllospadix*) and red algae cover the rocks. Lacking support, the exposed kelp *Laminaria dentigera* droops like a wilted lily. Surfgrass beds extend down to depths of 6 m. Not an algal seaweed, surfgrass is in fact a flowering seed-plant, an angiosperm, that is pollinated in the water. The flowers lack petals and nectar, of course—there is no need to attract bird or insect pollinators. The male and female flowers are on separate plants, with female plants outnumbering males 12 to 1 (Dawson 1966). A similar strong female-biased sex ratio also is true for an overlapping species, *P. torreyi* (S. Williams 1995). Surfgrass and the quiet-water eelgrass (*Zostera*) are the only submerged marine flowering plants in California. In contrast to algae, surfgrass and eelgrass absorb nutrients through roots as well as through their leaves (Terrados and Williams 1997). The long narrow leaves of surfgrass are only 2–3 mm wide, while eelgrass has leaves about 1 cm wide.

The seeds of surfgrass cannot attach directly to bare rock. Rather, they first adhere by barbs to established seaweeds such as small, articulated, coralline red algae, whereupon their growing roots secure

the plants to rock (Turner 1983). If an area of rock becomes denuded—by heavy surf or the impact of a floating log, for example—and there are surfgrass plants nearby, they will send in vegetative runners. Otherwise, newly opened space must first be populated by an alga. Once established, however, the surfgrass ultimately will dominate. Surfgrass itself has algae, snails, limpets, crustaceans, and other organisms living on it.

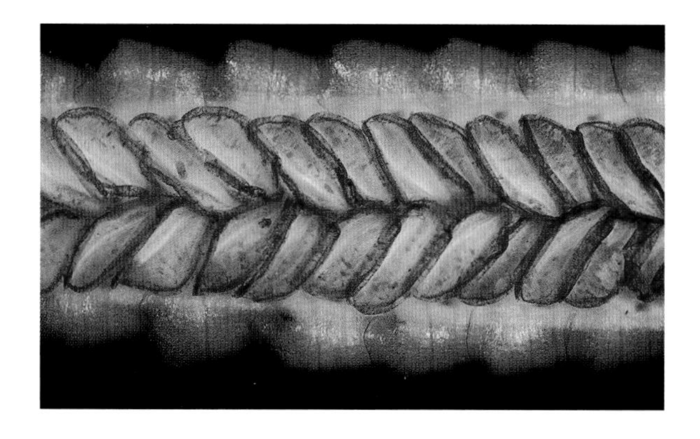

Phyllospadix with male flower, about 40 mm.
From low intertidal zone, Monterey Bay.

In this species of surfgrass the flowering shoots lie at the base of the leaves. This male flower shows double rows of pollen-producing anthers arranged like chevrons. Microscopic pollen is released in translucent strings from the sides of these anthers. The noodlelike pollen grains are 5 microns across and 1000 microns—a full mm—long. Pollination occurs when the "noodles" become entangled in little processes (stigmas) on the female flowers (Cox 1993).

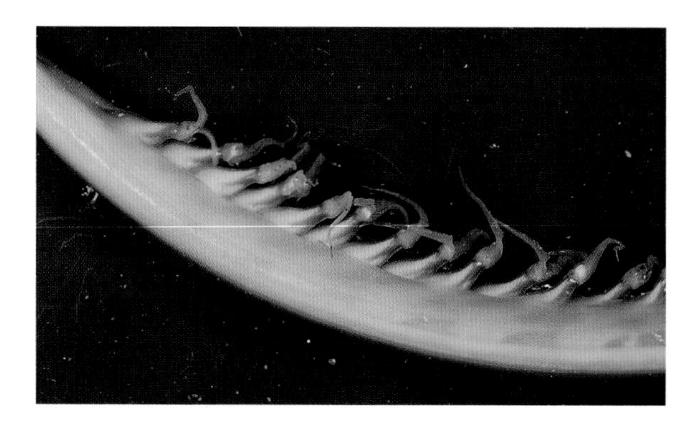

Phyllospadix seed pod, about 5 cm.
From low intertidal zone, Monterey Bay.

Note that each potential seed bud in the pod bears two bristly stigmas that trap the pollen "noodles." At the moment of contact, the surface materials on the stigma and on the pollen grain coalesce and bond in a manner apparently similar to that of epoxy cement (Pettitt, Ducker, and Knox 1981). Although pollination can occur underwater, the encounter is more likely when the seeds and pollen are floating together on the sea surface during very low tides, precisely the time that surfgrass often pollinates (Cox 1993).

Melobesia mediocris
Phylum Rhodophyta / Class Florideophyceae
and *Smithora naiadum*
Phylum Rhodophyta / Class Bangiophyceae

Opposite: Red algae growing on surfgrass *Phyllospadix scouleri,* leaf 3 mm wide.
From Carmel Bay.

Melobesia is an encrusting "coralline" red alga, forming a thin calcified crust on surfgrass and eelgrass—the only substrates it lives on. Only two thalli, or algal bodies, of the little seaweed appear in this photo, but often they are so numerous they coalesce into a continuous brittle mass. Conceptacles, sacs containing the reproductive structures, have tiny pores that open to the alga's surface. In this photograph, the conceptacles' positions are marked by dark spots. The alga with the smaller spots is a gametophyte; each of its conceptacles has a single pore. It would take microscopic examination of the conceptacles' contents to tell whether they are male or female, though males appear to be rare (Willcocks 1980). The alga with the larger spots is a sporophyte; its conceptacles have multiple pores.

Smithora is a foliose red alga, and, like the crustose *Melobesia*, it apparently grows only on surfgrass and eelgrass. The haploid blades arise annually by mitosis from flat, red, perennial cushions, well shown in the photo. These are young blades, but the blades often grow profusely to a length of 6 cm, arising especially at the edge of the host's leaves. *Smithora* blades mitotically produce sticky masses of spores in fertile parts called sori. The sori are released periodically at low tides, drift about, and finally adhere to other seagrass

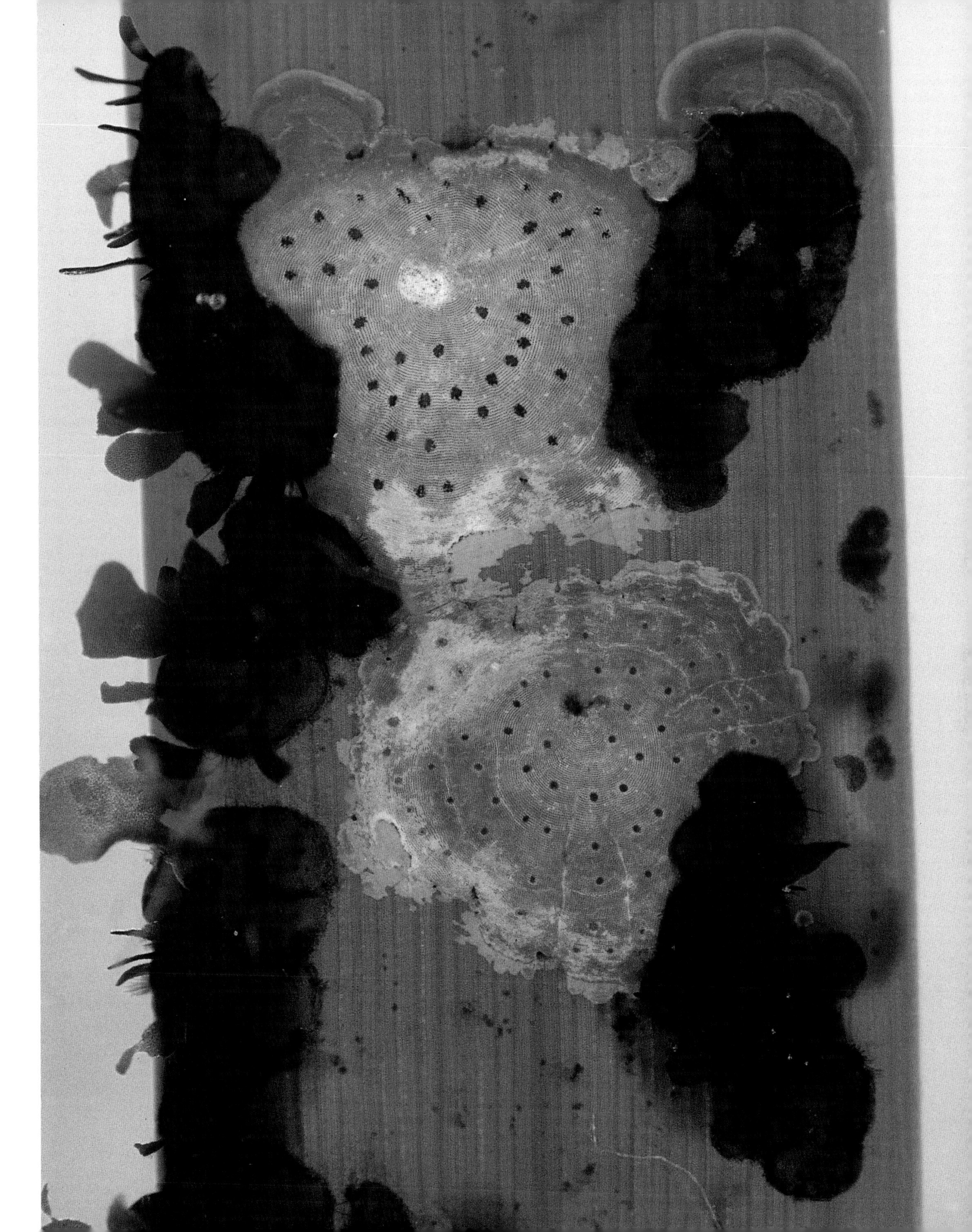

leaves, forming new basal cushions (I. Abbott and Hollenberg 1976)—essentially a cloning event. In a field study, the sticky sori quickly adhered and grew on artificial surfaces, and so it isn't clear why *Smithora* is so substratum-specific (Harlin 1973).

In addition to its prolific asexual reproduction by spores, *Smithora* can reproduce sexually. Both male and female gametes are produced on the same thallus, and then fertilization occurs (whether the alga is self-compatible remains unknown). Fertilization apparently is followed by the production of diploid carpospores, but here the trail ends. In the laboratory, culture of carpospores yields nothing further. Presumably they are released to produce an as yet undiscovered alternate phase in *Smithora*'s life cycle—perhaps a microscopic crustose or filamentous form. This unknown form may be what tides *Smithora* over during the winter, when many populations seem to disappear (Hawkes 1988).

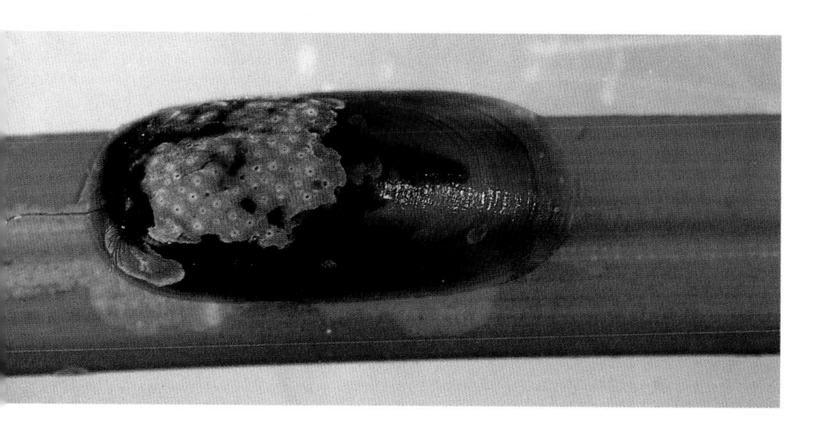

Tectura paleacea (= Notoacmea paleacea)

Phylum Mollusca / Class Gastropoda

Surfgrass limpet, 8 mm long, on *Phyllospadix scouleri*. From low intertidal zone, Monterey Bay.

These limpets are found only on surfgrass; at very low tides we see them in tidepools on the floating surfgrass leaves. The animal's narrow, elongated shape and its ability to adhere tenaciously to the leaves serve it well on this unstable grassy substratum. A small notch on the right side of the shell enables the limpet to circulate water through its gill even when it has withdrawn all its body parts within the shell to hang on in rough water. The surfgrass leaf in this photo shows a light green area with ragged edges; this is where the limpet has grazed the plant's chlorophyll-containing epidermal layer. Its feces are green and contain outer cortical material and chloroplasts from the surfgrass (Gansel 1979).

Note that the alga *Melobesia* is growing on this limpet's shell, perhaps providing it with camouflage. But *Melobesia* is a specialist that grows only on surfgrass, so what is it doing on this limpet? The answer: It "thinks" it's on surfgrass. *Tectura* picks up specific chemicals (flavonoids) while grazing surfgrass and incorporates them into its shell. Evidently the spores of *Melobesia* somehow recognize these chemicals (as do those of *Smithora*) and so are fooled into settling not on its typical host, but on the snail's shell (Fishlyn and Phillips 1980).

Incorporation of flavonoids in *Tectura*'s shell is a protective adaptation. The little tidepool seastar *Leptasterias* invades surfgrass beds and preys on snails there. Yet it has been observed to crawl right over this limpet, leaving it unharmed. For its part, *Tectura* simply clamps down tightly on its leaf of surfgrass and waits for the predator to pass. The limpet has been protected by chemical camouflage; the flavonoids in its shell have led the seastar—like *Melobesia*—to misidentify the shell as surfgrass. In contrast, two other snails that sometimes occur on surfgrass, *Lacuna marmorata* and *Alia carinata*, exhibit violent defensive actions when they detect the scent of approaching *Leptasterias*, for, not having flavonoids in their shells, they are often found and eaten by the seastar (Fishlyn and Phillips 1980).

Idotea montereyensis
Phylum Arthropoda / Class Crustacea

Isopod, about 6 mm long, on *Phyllospadix scouleri.* From low intertidal zone, Monterey Bay.

During most of the year these isopods live and graze on surfgrass in low intertidal areas with heavy surf. Their seven pairs of strong legs with spined tips help them cling tightly to the plant's leaves, and they gain some protection from predation by closely matching the leaves' color. In the spring the adults mate, and the mother broods her developing young in a pouch beneath her body. When the youngsters emerge, they find themselves in a very turbulent environment. Since their legs are not long enough to grasp the edges of the surfgrass leaves, they are swept inshore, where they take up residence in deep, relatively protected pools. There they feed on red algae and quickly change color to match their food source. At this time, too, through an unknown mechanism, the previous generation in these pools, now red adults, are stimulated to swim out to the surfgrass beds to forage. There, they quickly change to surfgrass green. These seasonal cross-migrations separate the young and the adult populations, enabling them to avoid direct competition for food and at the same time to achieve color camouflage.

Surprisingly, green *Idotea* are not green because they ingest an alga's green chlorophyll, and red *Idotea* are not red because they ingest an alga's red phycoerythrin pigment. How, then, do they accomplish their chameleon acts? The process is in fact fairly straightforward: The little herbivores ingest an alga's carotenoid pigments, modify them chemically, and deposit the appropriately colored pigment in their cuticle. They turn themselves red, for starters, by adding molecules of oxygen to an alga's beta carotene to produce the red pigment canthaxanthin. The next step, turning green, is a bit more of a challenge. First they bond the red canthaxanthin to a lipoprotein and reconfigure its protein molecule so it turns blue. Any artist can predict the next step: just add yellow. *Idotea* dissolves yellow carotene in the lipid portion of the blue lipoprotein, and green results (W. Lee 1966). The isopod's color changes can occur only coincident with molting.

In addition to these chemical tricks, the pigment granules in chromatophores beneath the cuticle, particularly the red ones, can expand or contract to fine-tune the animal's color match with its background. Red *Idotea* can concentrate its red pigment granules within thirty minutes, thus lightening its shade of red; dispersion of the granules to darken itself, however, is much slower. This physiological adjustment depends on the isopod's perception of light reflected from its substratum (Lee and Gilchrist 1972). The reddish isopod in this photo, though seen here on surfgrass, matches the color of the red algal epiphyte *Smithora naiadum*, which grows on surfgrass. We expect that with its next molt this *Idotea* turned green. Although the technique of color camouflage must provide some protection, even so, these isopods do make up a large part of the diet of tidepool fishes.

Idotea aculeata
Phylum Arthropoda / Class Crustacea

Isopod, about 10 mm long, on coralline red alga, *Bossiella* sp. From low intertidal zone, Carmel Bay.

Here we have a fantastic example of color camouflage: this little isopod blends in perfectly with the coralline alga where we found it. These herbivores also are found on surfgrass, but usually on plants encrusted with *Melobesia*, at which time they resemble that alga's pink hue. We do not know the precise pigments involved in these color changes, but the mechanisms revealed in *Idotea montereyensis* probably are a good guide for eventual investigations. Cryptically colored though it is, this isopod, too, has been found in the stomachs of tidepool fishes.

Serpulorbis squamigerus
Phylum Mollusca / Class Gastropoda

Serpulorbis snails on piling, 10 mm. Municipal Wharf II, Monterey.

The calcareous tubes of these "unrolled" snails, about 10 mm across and fixed to the substratum, superficially resemble the lime casings of some big worms. But these tubes have the same structural layers as a snail shell, and the body within is clearly that of a snail, not a worm. During growth this snail's shell, rather than taking on the usual gastropod coil, forms a meandering tube, cemented in place. Here, surrounded by erect bryozoans, the entrances of some of the snails' tubes barely show their inhabitants' heads. The now-immobilized snail feeds by secreting copious amounts of mucus from a foot gland to form a veil-like net that it holds with its tentacles. The net traps small plankton and detritus. Several snails may form a communal net in a behavior reminiscent of communal net-spinning by social spiders. After filtering for half an hour, *Serpulorbis* reels its net in and ingests it, trapped food and all.

These snails are usually sequential hermaphrodites, initially male and later female. Males release packets of sperm (spermatophores) into the plankton, which may become entangled in a female's mucus net. There, activated by the female's feeding movements, the spermatophores ejaculate sperm that enter the female (Hadfield and Hopper 1980). Sperm can be stored by the female for long periods until she becomes reproductive. Like the subtidal sessile snail *Petaloconchus montereyensis* (see chapter 5), *Serpulorbis* produces two types of sperm, one nucleated and sexually active and another that lacks a nucleus and serves only to nourish "normal" sperm. Fertilized eggs attach in masses inside the mother's tube to become brooded embryos. After release and life in the plankton, the larvae eventually settle gregariously to form new clusters of sessile adults.

Until recently, this species was rare as far north as Monterey Bay (Barry et al. 1995). Today large masses of *Serpulorbis* thrive here, particularly on low intertidal rocks in Monterey Harbor. We don't know why they suddenly have proliferated, but their abundance may be due to a long-term warming trend in our chilly coastal waters.

Serpulorbis snail, about 5 cm long. Carmel Bay, 12 m deep.

Although *Serpulorbis* usually settles gregariously, solitary ones like this snail may reproduce successfully, since a male's spermatophores can be borne in the plankton to females as much as a meter distant (Hadfield and Hopper 1980). This one's tube shows the species' characteristic ribbing. It is surrounded by the cobalt blue sponge, *Hymenamphiastra cyanocrypta,* which derives its intense color from symbiotic bacteria.

Anthopleura elegantissima

Phylum Cnidaria / Class Anthozoa

Aggregating anemones at low tide, each about 30 mm. Intertidal rocks, Carmel Bay.

These small mid to low intertidal anemones spread all over the rocks. They're easily missed because the squishy little bodies are mostly covered with bits of shell and coarse sand, so that when they close up at low tide they become inconspicuous. The intertidal zone is a stressful environment, particularly when the tide is out, but a jacket of sand gives these animals some protection from ultraviolet light and from desiccation. Like several other local anemones and like the tropical reef-building stony corals, *Anthopleura* harbors symbiotic, single-celled algae within its cells. From the algae's photosynthetic food-production *Anthopleura* can obtain much of its own nourishment (Muscatine and Hand 1958), especially when small crustaceans are scarce. *Anthopleura* is also able to withstand long periods of burial under sand; like sprinters, the anemones can burn carbohydrates anaerobically and pay off their oxygen debt when conditions improve.

Anthopleura elegantissima was discovered in the 1820s at Sitka, Alaska (then capital of Russian America), during an expedition sent by Czar Nicholas I. Specimens of the anemone were brought back to the

Imperial Academy of Sciences in St. Petersburg, where they were described and named "most elegant" by the famous German naturalist J. F. Brandt in 1835. The original account of the expedition, published in Paris (Postels 1836), makes fascinating reading even today. It conveys the sense of wonder and dedication of these naturalist-explorers and their steadfastness under ferocious conditions in the field.

Above left: Oral disc, *Anthopleura sola,* about 10 cm. Piling, Municipal Wharf II, Monterey.

Anthopleura occurs in both cloning and solitary forms (McFadden et al. 1997). Each form reproduces sexually, but the solitary one, *Anthopleura sola,* does not clone.

When immersed at high tide, an expanded *Anthopleura sola* reveals its oral disc with its central mouth. The anemones' radial symmetry and diffuse nerve net facilitate the collection of food and the recognition of predators coming from any direction; they are well adapted for a lie-in-wait strategy. The tentacles are loaded with tiny venomous stinging capsules (nematocysts) with which they capture prey. (See discussion of *Paracyathus stearnsii,* chapter 5.)

Anthopleura elegantissima cloning by division, field of view about 20 cm. Piling, Municipal Wharf II, Monterey.

When reproducing asexually, the aggregating anemone stretches—as seen in this photo—and literally pulls itself apart to become a genetically identical pair. This is done repeatedly to form vast clones of individuals alike in sex and color. Under favorable conditions, rapid replication soon enables well-adapted individuals to dominate space in their habitat.

We found these *Anthopleura elegantissima* anemones living on shaded pilings; in contrast to their intertidal brethren, they seem to thrive without a sandy-shelly coating. They have retained their symbiotic algae, though, as shown by their dark color. *Anthopleura* also reproduces sexually: male polyps release sperm, female polyps release eggs, and fertilization occurs in the sea. The resulting planktonic larvae, known as planulae, disperse and eventually settle and metamorphose into polyps that start new clones.

Warfare between *Anthopleura elegantissima* clones, field of view about 25 cm. Piling, Municipal Wharf II, Monterey.

At the touch of a tentacle, individuals in one clone can recognize as non-self those in an adjacent, genetically different clone, and they will trigger an aggressive fight (Francis 1976). Pretty amazing for a creature without a brain! As shown in our photograph, a no-man's-land between the two clones becomes established, lined by specialized warrior polyps. When stimulated, the warriors inflate white swollen bulges that look like stubby tentacles (acrorhagi), located just beneath the feeding tentacles. These are loaded with nematocysts and used as weapons during interclonal fights. When applied to the victim, acrorhagi can inflict severe injury or even kill polyps that do not retreat from the battle line. It's a battle to dominate space for one's own genes. Warrior polyps are not active sexually, they are smaller than sexually reproductive polyps, and they have more acrorhagi. Look closely, and you can pick out warriors in our photo.

During these aggressive encounters, the acrorhagial response is no greater or faster with a second or third encounter than it was with the first one. In other words, there is no evidence of specific immunologic memory in these simple animals: the clones do not learn (Lubbock 1980). It's an important distinction, as immunologic memory is a hallmark of our own enormously complex immune system. That's why vaccines work.

Anthopleura elegantissima and the plumose anemone, *Metridium senile,* field of view about 25 cm. Piling, Municipal Wharf II, Monterey.

Anthopleura's reaction to injury. Piling, Municipal Wharf II, Monterey.

Here we see two anemone species facing off in a manner reminiscent of *Anthopleura*'s interclonal fights. *Anthopleura* will fight with its acrorhagi, while *Metridium* has its own specialized weapons, transformed feeding tentacles called "catch tentacles" that are loaded with nematocysts. Note one very exposed little *Anthopleura* combatant in the middle of the no-man's-land, its tentacles withdrawn, its acrorhagi expanded, defending itself. It may well succumb if it does not retreat from the firing line.

Remarkably, there are subtle differences in the way *Anthopleura* uses its weapons. It will deploy its acrorhagi only against anemones, but its feeding tentacles are much less specialized—those nematocysts will readily discharge following contact with a variety of prey and predators (Lubbock 1980; Lubbock and Shelton 1981).

You might not expect it of an anemone, but even in the absence of vocal cords *Anthopleura elegantissima* can warn its neighbors of danger. Here Libby has firmly struck a polyp located just beneath the field of view, and a few seconds later, as can be seen in Lovell's photograph, a characteristic wave of response has spread to its neighbors: they outwardly flex their tentacles, touching the column, and then retract and close up, completely covering the tentacles and oral disc. The injured anemone has released a chemical signal, an alarm pheromone, to which nearby animals are reacting (Howe and Sheikh 1975). This reaction contrasts with the feeding response, initiated by an amino acid from the prey's protein, that induces the receptive bending of the tentacles that brings prey to the mouth (Lindstedt 1971). For *Anthopleura,* satisfying hunger may be more important than avoiding pain: should it receive conflicting stimuli both for feeding and for alarm, the feeding response dominates (Howe 1976).

10

2 Wharfs and Docks

The floating docks of Monterey Harbor and the pilings beneath Municipal Wharf II provide a rich and diverse community created largely by human activity. Such communities have imaginatively been called "accidental zoos." Over the centuries ships have traveled from port to port, including Monterey, dumping ballast water containing whole communities of larvae from distant ports, and their moored hulls are often encrusted with foreign invertebrates that may be reproductive. A tunicate we discuss, for example, *Botryllus schlosseri,* is probably an immigrant from northern Europe. It was reported in the northeastern United States in 1903, in New Zealand and Australia by 1928, in San Francisco Bay in the 1940s, and in Washington State in the late 1960s (Lambert and Lambert 1998); today it's common on the docks in Monterey. The bryozoan *Bugula neritina* is likewise widely dispersed. These and many other species now are cosmopolitan, appearing in harbors—often only in harbors—worldwide.

At low tide the pilings, shaded by the wharf, form an intertidal zone without algae and without herbivores. Familiar, no; informative, yes! As revealed in the photo, zonation is sharply delineated. The mussel/barnacle zone is out of water much of the time. A little lower, in a zone immersed most of the time, live aggregating anemones without their usual shelly coating and now partially closed. Still lower and approaching the low-tide level, a flash of red reveals strawberry anemones, usually subtidal organisms that only infrequently are exposed to air. This community is spared many of the vicissitudes of rocky intertidal life—no burning sun and UV light, no pounding surf, little osmotic stress from direct rain. The sessile organisms depicted here are all relatively young; these concrete pilings had recently replaced earlier wooden ones that were riddled with shipworms (actually a kind of clam) and gribbles (crustaceans).

Subtidally, the pilings are a treasure trove of biological interactions. Huge ochre seastars engorge themselves with mussels. Nudibranchs cluster about their anemone

Pilings at low tide. Municipal Wharf II, Monterey.

prey. Anemone clones, as we saw in chapter 1, do battle across strictly demarcated no-man's-lands. And here we find unusual organisms such as phoronids, enigmatic suspension feeders that we have not seen elsewhere along Monterey's shores.

A wharf is a spooky place to dive. It's dark, with visibility often only an arm's length. Fishing lines threaten entanglement, and raucous sea lions flash by. But we have a strong incentive to keep returning to this distinctive habitat, where something new seems to lurk behind each piling.

The nearby docks of a marina, unlike the wharf's pilings, float up and down with the tides. They present the odd ecological situation of a subtidal habitat that is immediately accessible: all you need to do is sprawl on the dock and peer over the side. Like the pilings, in many ways it's a benign habitat. But a lot goes on in a restricted area, and it is home to many fascinating, tiny, fragile beasts: baby skeleton shrimps crawl over their mother; sea spiders repose on hydroids; compound tunicates overgrow seaweeds and invade each other; and cruising amphipods maneuver their way through it all. When we record the tiny, living subjects in action under magnification in our lab, we gain insights that otherwise would escape us.

Aglaophenia latirostris
Phylum Cnidaria / Class Hydrozoa

Ostrich plume hydroid, about 8 cm. Carmel Bay, 14 m deep.

If you lie prone with your head hanging over the edge of a floating dock, you may be only inches from the bouquet of little amber feathers of a hydroid colony like this one. A carnivorous animal, it dines on extremely tiny prey. Hundreds of feeding polyps, each with a crown of tentacles around the mouth, line the branches of its delicate pinnate fronds. The tentacles are loaded with nematocysts, stinging cells with harpoonlike threads, that it uses to immobilize its prey. This colony is swarming with tiny, inconspicuous "skeleton shrimp," caprellid amphipods. Although the scene was photographed in Carmel Bay, the species is common throughout the protected waters of Monterey Harbor.

Most hydrozoans have two very different stages in their life cycle, a polyp stage and a medusa stage. These plumes characterize the hydroid's sessile polyp stage. A founding polyp (either male or female) buds off new polyps to form this feathery colony. Individual polyps are differentiated and serve separate functions for the benefit of the entire colony—feeding, reproduction, or defense. The reproductive polyps in hydroids enlarge and produce the life cycle's next phase. In many species this consists of little free-swimming hydromedusae (jellyfish). In others, such as *Aglaophenia*, this stage is retained on the parental colony and the sessile "medusoids" reach sexual maturity while attached there. Such complex hydrozoan life cycles inspired Ed Ricketts to write that one might as well be "asked to believe that rose bushes give birth to hummingbirds, and that the hummingbirds' progeny become rose bushes again" (Ricketts and Calvin 1985).

Below: Colony of *Aglaophenia latirostris* with extended feeding polyps and corbula, about 4 mm. From floating dock, Monterey Harbor.

This magnified view shows the tiny feeding polyps of *Aglaophenia* with their extended tentacles, which can trap even the tiniest zooplankton. Surrounding the tentacles of each feeding polyp are small defensive polyps, especially rich in stinging cells.

In *Aglaophenia* the reproductive polyps are enclosed in large, swollen, orange structures called corbulae, one of which is seen here. The medusoids of this species, which remain attached, grow inside the colony's many corbulae. Male medusoids shed sperm into the sea that fertilize eggs in the medusoids of female colonies. The *Aglaophenia* zygotes develop into tiny, ciliated, crawl-away larvae. These soon settle and metamorphose to become the first polyp of a new feathery colony.

Aglaophenia latirostris is completely harmless to people, however devastating it may be to microscopic zooplankton. But in the tropical Pacific a relative in the same genus, *A. cupressina,* has equally small polyps that deliver a fearful sting—as we can attest. Why should one species be so benign and the other so ferociously toxic?

Tanystylum duospinum

Phylum Arthropoda / Class Pycnogonida

Opposite: Female sea spider, 8 mm across, on hydroid *Aglaophenia latirostris.* From low intertidal zone, Monterey Harbor.

The exoskeleton and the jointed legs of this tiny animal reveal its phylum-level arthropod identity, but who are its closest relatives? Interestingly, the six hundred species of sea spiders (pycnogonids) are often placed in the subphylum Chelicerata, which includes land spiders, scorpions, mites, ticks, and horseshoe crabs.

The small size, slow and deliberate movements, and cryptic form and habits of sea spiders make them a challenge to find in nature. In the laboratory, we often spot *Tanystylum* on the fronds of the hydroid *Aglaophenia latirostris*, on which it feeds. This one has lost one leg. Unlike crustaceans, most adult pycnogonids do not molt and so cannot replace lost parts. The rather swollen legs suggest that this is a female; the gonads and often the yolky eggs themselves fill a pycnogonid's legs, for lack of space in its tiny central body.

The legs of sea spiders, like those of land spiders, lack extensor muscles. The legs, despite being flexed by muscles, can be extended only by the hydraulic pressure of body fluids injected into them. These animals are quite sluggish, though agile. Their hydraulic leg extension cannot account for their sluggishness, however, for on land jumping spiders leap about athletically, and many other spiders run rapidly by the same combination of hydraulics and musculature.

Below: Male *Tanystylum* bearing clusters of eggs, 8 mm across, ventral side, on *Aglaophenia.* From low intertidal zone, Monterey Harbor.

During mating, the male *Tanystylum* clings beneath the female with their ventral surfaces touching and their heads lying in opposite directions. Fertilization occurs as eggs and sperm are emitted from pores in the animals' legs. Then the male takes over the joys of parenthood. Here, a male is brooding the eggs he has gathered from his mate and attached to two specialized legs, where the distinctive larvae will develop. Later during their development, these larvae will parasitize *Aglaophenia* (Russel and Hedgpeth 1990).

Although isopods, anemones, and fishes reputedly prey on pycnogonids, one sea spider (*Pycnogonum literale*) neverthless has thousands of defensive epidermal glands scattered over most of its body, each associated with a sensory hair. Following the hair's stimulation, the glands release noxious secretions (ecdysteroids) as a chemical defense against predators. Such pores occur in many pycnogonids; chemical defenses may be common to all sea spiders (Tomaschko 1995).

Many other features of the physiology and life history of these anatomically bizarre invertebrates remain to be studied by the very few specialists who have taken on this group.

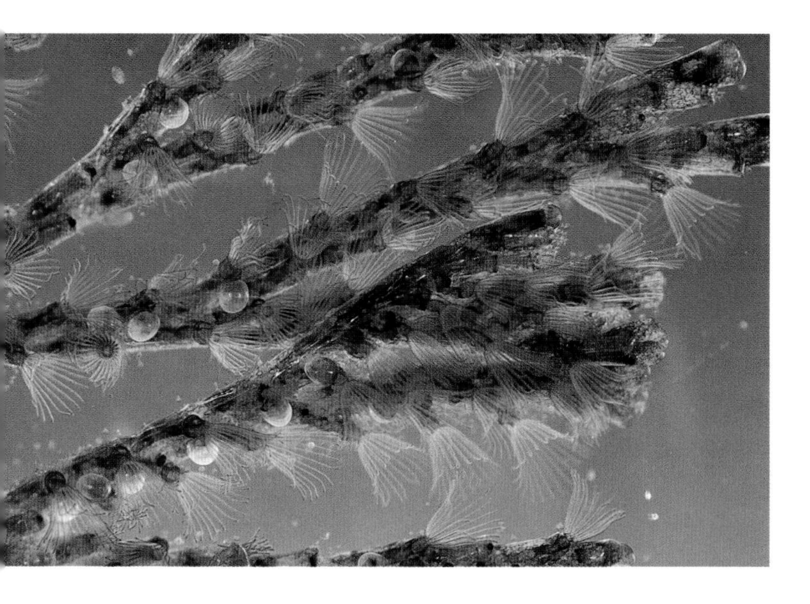

Bugula neritina
Phylum Bryozoa / Class Gymnolaemata

Branching bryozoan, field of view about 6 mm. From floating dock, Monterey Harbor.

This species of *Bugula* is a well-known cosmopolitan fouling organism—that is, one that settles on boat hulls, pilings, floats, and other artificial substrata. The colonies grow as little purplish "bushes" about 30 mm high. Bryozoans are colonial animals made up of hundreds of small (<1 mm) filter-feeding individuals, or zooids. The colony begins when a larva settles, metamorphoses, and buds; subsequent generations of budded zooids form an upright colony. Each zooid has a crown of ciliated tentacles, called a lophophore, that it extends for feeding. The cilia drive a current toward the centrally located mouth. When feeding on zooplankton, *Bugula* forms its lophophore's tentacles into a cage with the tops of the tentacles twisted tightly together, and thus traps active prey (Winston 1978). For most bryozoans, the ciliary current alone suffices to trap minute phytoplankton.

Bryozoan zooids are hermaphroditic. Sperm apparently emerge from pores at the tips of specialized tentacles and somehow gain access to the ova inside other zooids, allowing cross-fertilization to occur between colonies. One yolky egg at a time is brooded in a translucent, globular ovicell, several of which are conspicuous in this photo. Within the ovicell, the developing embryo feeds on the egg's yolk and on material delivered from the maternal colony by a placenta-like system. Could the two small yellow bodies in the upper left be escaping yolk-laden larvae? After the larvae have slipped out of the ovicells, they swim for only a few hours and then settle. Usually they settle and metamorphose only next to sibling larvae, not next to adults or unrelated larvae, even those of their own species (Keough 1984), which means that somehow they recognize their kin. Such larval kin-recognition occurs in a few vertebrates, but it is not known among other invertebrate larvae (though many do seek out the company of adults of their own species when settling).

These relatively conspicuous larvae may avoid fish predation by spending only a short time in the plankton, but they are also chemically defended (Lindquist and Hay 1996). In one experiment, fringed file fish from the Carolinas attacked only a few *Bugula* larvae, and those larvae that were attacked were quickly rejected and went on to successful metamorphosis. The fish also rejected all food pellets treated with larval "extract," but they readily consumed pellets treated with extracts of settled *Bugula* colonies, which suggests that the adults are chemically undefended. Interestingly, compounds (bryostatins) extracted from colonies of *Bugula neritina* are potent anticancer agents and stimulate the growth of human immune cells (Patella et al. 1995).

Bryozoans are in a distinct phylum, Bryozoa, and with the phyla Phoronida and Brachiopoda can be considered together in one sprawling group, the Lophophorates, since they all utilize a similar tentacular feeding organ.

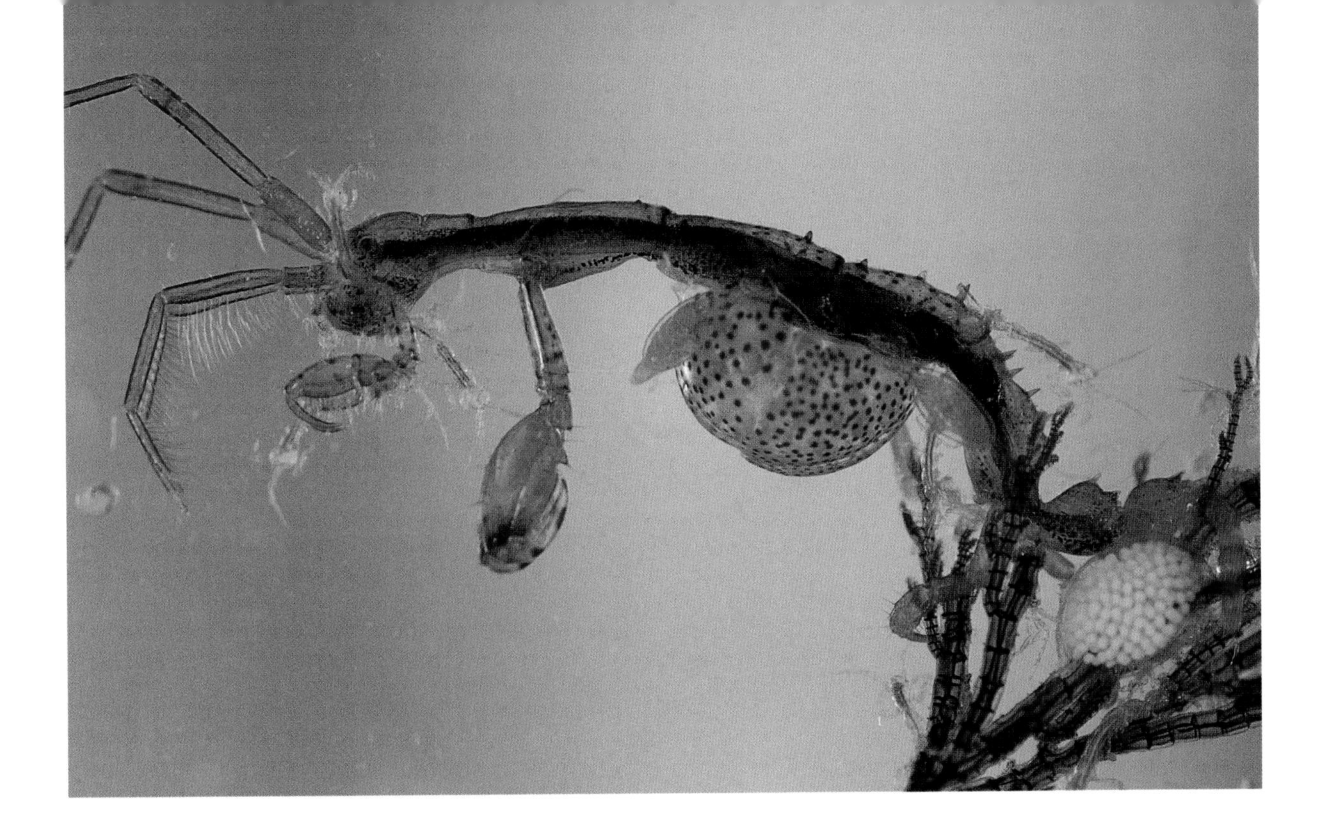

Caprella sp.
Phylum Arthropoda / Class Crustacea

Female "skeleton shrimp," about 12 mm long, on filamentous red alga. From floating dock, Monterey Harbor.

If you peer closely at a clump of algae, erect bryozoans, or hydroids, you frequently find it crawling with bizarre amphipod crustaceans called caprellids or skeleton shrimp. Caprellids have a greatly elongated thorax and a reduced abdomen devoid of appendages. The female caprellid in this photograph is clinging to an alga with her posterior three pairs of thoracic legs. Thus she is free to use her ferocious, clawed anterior two pairs of thoracic limbs to capture food, spar with other caprellids, and groom herself. Her fringed second antennae are used for swimming and for filter-feeding. Little paddle-like gills and a red-dotted marsupium (brood pouch) that protects her brooding young arise from her fourth and fifth thoracic segments. The brooded eggs develop directly into juveniles, five of which are here climbing over their mother. Although the young of some caprellid species disperse immediately after emerging from their mother's brood pouch (Aoki 1997), the tiny, fragile young of this species demonstrably turn to their mother for care. Among the major predators of caprellids are surfperch, which usually take larger individuals that are actively swimming, crawling, or feeding. (In this photo, the cluster of eggs attached to the alga appears to be that of a mollusc.)

Caprellids climb about by securing themselves first with their anterior legs and then with their posterior ones, like agile inchworms. They scrape diatoms and nutritious bits of debris from nearby surfaces and filter small food particles from the water while bowing and waving their feeding arms.

When reproductive, the male caprellid guards his mate, grasping her with his anterior thoracic appendages. He will aggressively defend her from other covetous males in often mortal combat. When the female is ready to molt, he removes her exoskeleton, they copulate, and she deposits her eggs in her brood pouch (Caine 1991).

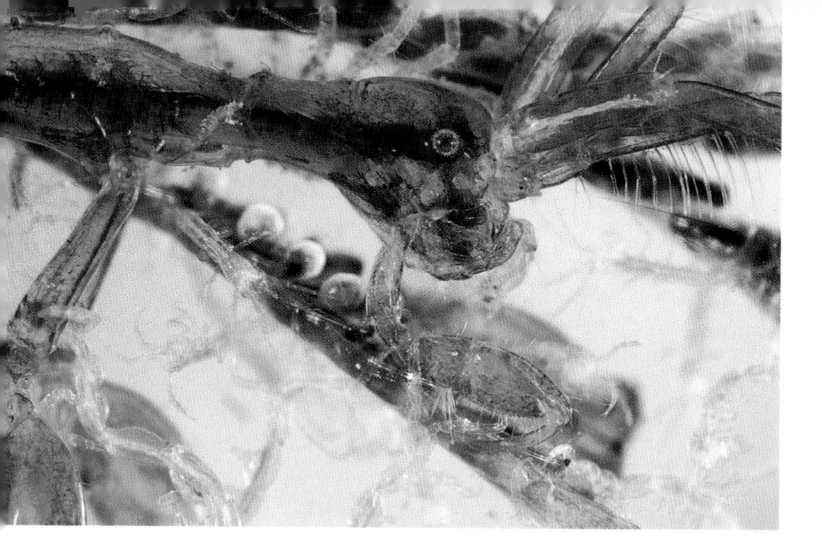

Caprella on the bryozoan *Bugula neritina,* 4 mm. From floating dock, Monterey Harbor.

This skeleton shrimp does a good job of matching its bryozoan background. Many caprellids have chromatophores that concentrate or disperse intracellular pigments to produce rapid color changes. How these animals "decide" on this delicate cellular adjustment to match their surroundings remains deeply puzzling, but it probably is determined by the amount and quality of light entering their eyes.

Jassa, probably *marmorata*

Phylum Arthropoda / Class Crustacea

Opposite: "Cruising male" amphipod, about 12 mm long, on a red alga. From floating dock, Monterey Harbor.

Jassa marmorata, a nonnative species, is often found in our harbors; like so many invasive marine invertebrates, it spread by shipping. Amphipods in this genus have remarkable anatomical differences between the sexes in addition to their behavioral differences. Look closely at the conspicuously broad appendage, the second gnathopod, or feeding arm, of this male. It is twice the size of the corresponding part in females of the same species, and it has a large, thumblike projection that is absent in females and younger males. The "thumb" forms only with the final molt, when males stop growing. Amphipods use their gnathopods to manipulate food, but clearly this odd, nonprehensile appendage must serve some other purpose. Careful study in the laboratory has revealed that indeed it does.

These amphipods live singly in tubes constructed of debris glued together with mucus secretions. They feed on algae and on particles trapped by their antennae. Females molt throughout their lives and, when mature, ovulate after each molt. To juvenile males the gnathopod's thumb signifies adult dominance, and to females it signifies mating intent (Borowsky 1985; Conlan 1989). Thumbed, sexually mature males leave the safety of their tubes and cruise about seeking receptive females. When the female is about to molt, the male will be waiting on her tube. Copulation follows, and the female broods the fertilized eggs and directly developing young. Males use their gnathopods to spar with each other for access to females. The males with the largest thumbs usually

prevail, and females show a clear preference for them. After copulation, the aptly described "cruising males" move on, seeking other receptive females.

Females and young males seldom leave their tubes. Cruising males, of course, are dangerously exposed to predatory fish, but only by leaving the safety of their homes do they have the opportunity to mate with many females. The way these male amphipods use their gnathopods to spar with each other brings to mind the contests that occur between elk, rhinoceros beetles, and many birds to gain females' sexual favors.

Botryllus schlosseri
Phylum Chordata / Subphylum Tunicata / Class Ascidiacea

Opposite: Compound tunicate showing a "system" with "takeover," 5 mm across. From floating dock, Monterey Harbor.

This is one of our favorite animals: it's beautiful, it's sophisticated, and as a chordate it's a relative of ours! It wasn't until 1858, however, that Alexander Kowalevsky, a Russian academician of St. Petersburg, noted that tunicate larvae, which resemble tiny tadpoles, have the distinctive traits of a vertebrate embryo (Fig. 6). Specifically, they have gill slits, a dorsal tubular nerve cord, a notochord (the stiffening rod in vertebrates that precedes the development of the spinal column), a post-anal tail, and a general resemblance to the vertebrate body plan. Although the "tadpole's" tail, nerve tube, and notochord disappear with metamorphosis, the gill slits divide repeatedly to form an elaborate perforated pharynx: the adult's mucus-covered filter-feeding apparatus.

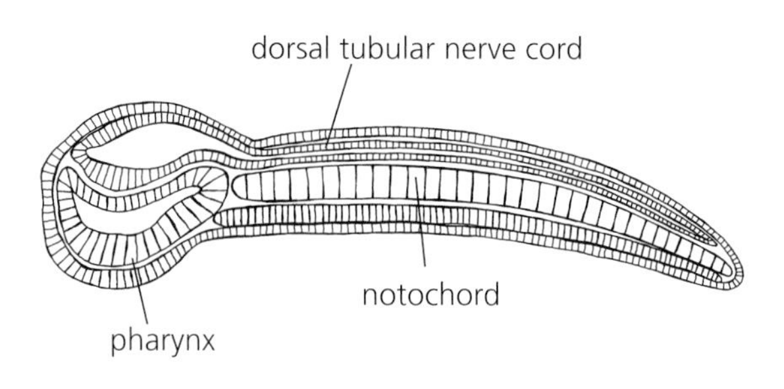

Fig. 6. "Tadpole" larva (after Kowalevsky 1866, in Hyman 1922)

Today, over a century after Kowalevsky's prescient observations, evidence continues to mount indicating that tunicate larvae probably are surviving examples of the earliest chordate body plan from which vertebrates arose (Whittaker 1997). An analysis of the RNA of ribosomes (tiny intracellular organelles involved in protein synthesis) in tunicates and in vertebrates, for example, suggests that they indeed are sister groups. So both by anatomical traits and by our genes there seems little doubt that we ourselves are related to the sea squirts. In what follows we present several examples of these hard-to-believe creatures.

We have spent a lot of time lying on a floating dock peering over the side with our arms plunged into seawater almost to the shoulder, retrieving tiny bits of bright orange on blades of seaweed—colonies of *Botryllus.*

This flowerlike "system" consists of a number of zooids—the flower's "petals"—that have arisen by budding from earlier zooids. Each zooid, or individual in the colonial organism, has an oral water-intake pore; together they share the system's central common outlet pore. During growth, systems of zooids reorganize and multiply within the colony as a whole. Each zooid in the system lives for about seven days and then dies by a process of "programmed cell death" known as apoptosis. Zooids bud off new ones laterally, and as an old zooid degenerates its newly budded progeny commence feeding. Eggs, which develop slowly, move through blood vessels to a succeeding generation of petals, until they are ready for fertilization and tadpole release. The whole colony shows synchronized waves of this replacement of zooids by their offspring, a phenomenon called "takeover." In the system shown here, the senescent zooids are orange and the young ones, already with open oral pores, are yellow. The surrounding little orange blobs are pigment-containing cells in blood vessels that pervade the supporting, transparent, gelatinous covering (the tunic).

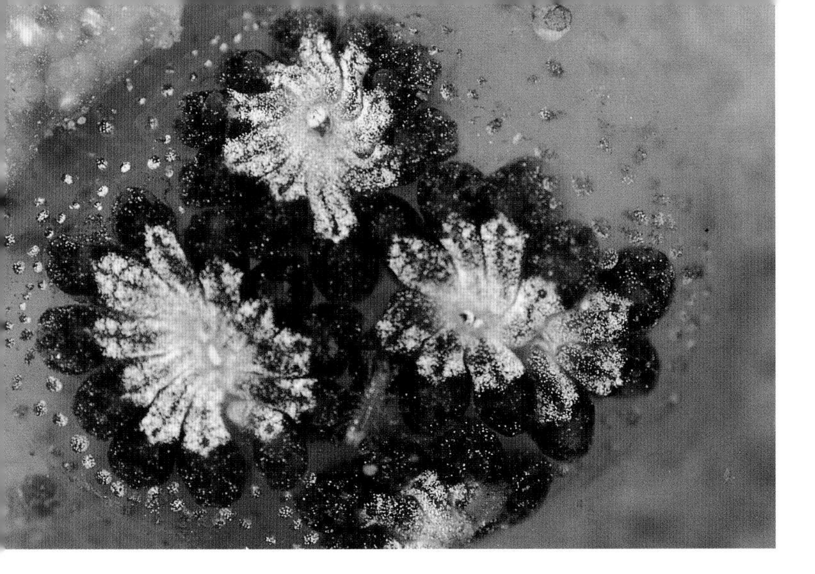

Botryllus colony, about 12 mm across, growing on another species of compound tunicate. From floating dock, Monterey Harbor.

A colony like this young one at first had only a single system of zooids, but it now has grown into four systems of genetically identical zooids, all the same color, all hermaphroditic, all the progeny of a single founding larva. These cloned animals inherit a genetically programmed metabolic clock with the same setting: all of the zooids, even those that are physically separate, bud, release their larvae, and undergo takeover in lockstep with their siblings.

Entire *Botryllus* colonies have a programmed life span, just as their zooids do. Field populations of this species, whether from Monterey Bay, from the Venetian lagoon, or from the Mediterranean coast of Israel, all live for about three months (Rinkevich et al. 1992). The seven-day life span of a system's zooids and the three-month life span of a whole colony appear to be under separate genetic control. What causes death in these animals is not at all clear, but one thing is sure: it is far from a fortuitous accumulation of accidents.

In all multicellular animals, including *Botryllus* as we have seen, apoptosis (genetically programmed cell death) serves to weed out superfluous or disordered cells. Our own aging and death appear to combine progressive "wear and tear" and genetically programmed intrinsic processes. The failure of apoptosis may play a role in human cancer formation, when cells proliferate wildly and do not self-destruct (Duke, Ojcius, and Young 1996), and too little apoptosis could play a role in autoimmune disease. In contrast, too much apoptosis may increase damage following strokes and the neurodegeneration of Alzheimer's disease (Miller 1998). Currently, the signaling pathways that tell cells when to die are being intensively studied.

Botryllus colonies come in many colors, frequently orange. This purple colony's silvery granules (metabolic products) lie in blood cells; those in the surrounding tunic demarcate blood vessels connecting the systems.

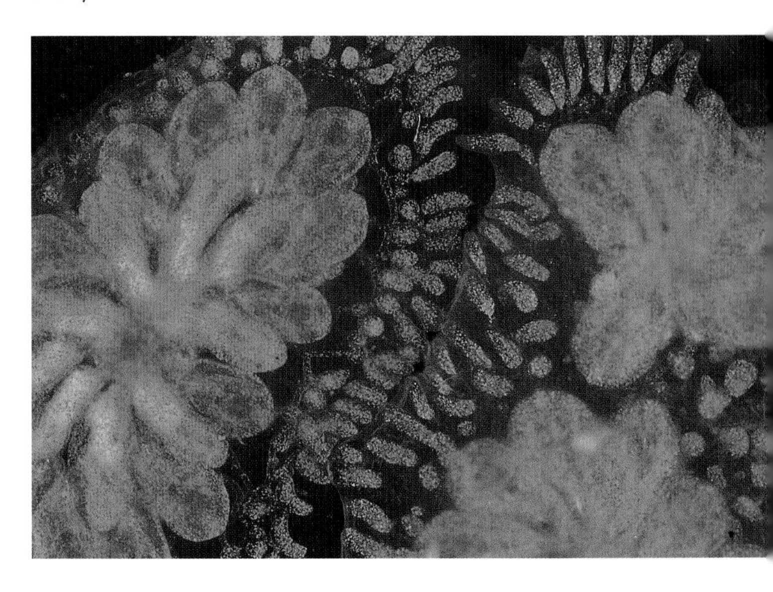

Botryllus colonies undergoing rejection, about 15 mm across. (Specimen courtesy of Kathi Ishizuka, Hopkins Marine Station, Stanford University)

Although transplantation of organs only occurs through surgical intervention in humans, in these protochordate tunicates a very similar phenomenon occurs naturally. *Botryllus* colonies recognize "self" and non-self, and adjacent colonies will fuse or reject each other according to the dictates of a single Mendelian "fusibility gene." In simple terms, genes determine proteins. When colonies meet, the animals somehow compare their cells' surface proteins. If these are similar, the colonies consider themselves "self" and fuse; if the proteins are dissimilar, as in this example, rejection sets in with clotting in the blood vessels, death of tissue (the dark brown material), and separation of the colonies' tunics. Cells resembling our own antibody-producing white blood cells appear at the site of the colony's immune reaction.

Such a rejection reaction suggests the phenomenon of organ transplantation failure in humans. Tunicates are just close enough to our evolutionary lineage that they help us to understand why such failures occur and what genetic signals regulate transplant compatibility (Scofield, Schlumpberger, and Weissman 1982; Saito, Hirose, and Watanabe 1994).

Right: Fusion between colonies, about 15 mm across. (Specimen courtesy of Kathi Ishizuka)

These colonies have been bred in the laboratory, so their family history is known. Sharing the same surface proteins, they recognize each other as "self." Note that blood vessels are growing between the fused colonies; they establish a common circulation, identified by the vessels' yellow-pigmented blood cells. Undifferentiated cells will crisscross through the vessels, giving rise to new generations of cells that will help form new zooids. Gradually, a colony will emerge that is a mixture of cells from the two donor colonies. Such a mix is called a chimera. Should cells with genes coding for the physical traits of one colony (e.g., color) dominate, the chimera will come to resemble that parent. But germ cells from the other colony—for example, those that produce sperm—may flourish independently and dominate in the same chimera. The colony's appearance thus may mask the winner of a subtler contest; this winner will pass on its genes to the colony's sexual progeny (Stoner and Weissman 1996).

If the concept of a chimera is unfamiliar to you, recall that the Chimera of Greek mythology was a fire-breathing she-monster with a lion's head, a goat's body, and a serpent's tail—different beasts combined into one.

The concept of chimerism in *Botryllus* may be more relevant to us humans than one might suppose. Humans become chimeras after organ transplantation, and serious autoimmune disease may follow in the host. Chimerism also may occur during pregnancy when a few fetal cells escape into the maternal circulation and establish their own cell lines; as evidence, many years after childbirth, minute amounts of male DNA can be found in some cell populations in women who have given birth to sons. The issue is of great importance, as these persistent fetal cell lines could in some instances be responsible for autoimmune disease. The body then attacks its own tissues—a reaction to non-self (J. Nelson 1998).

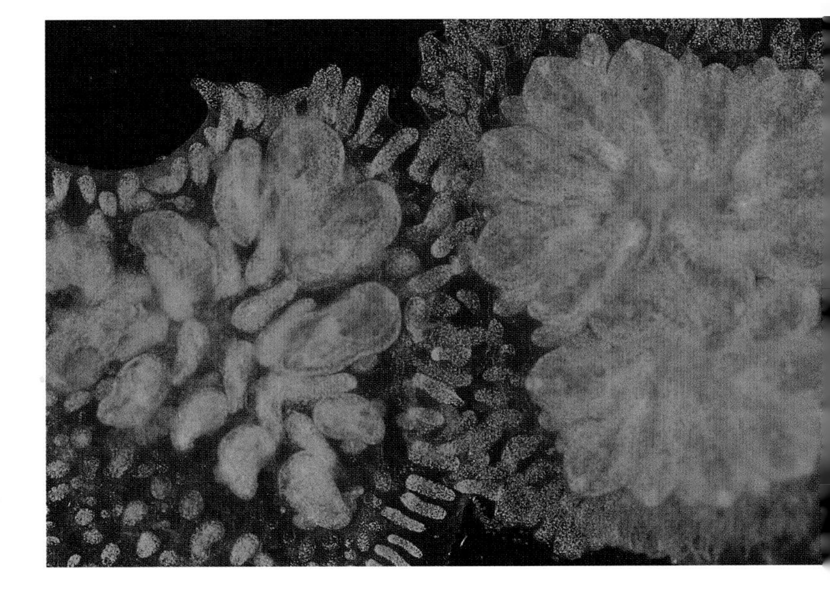

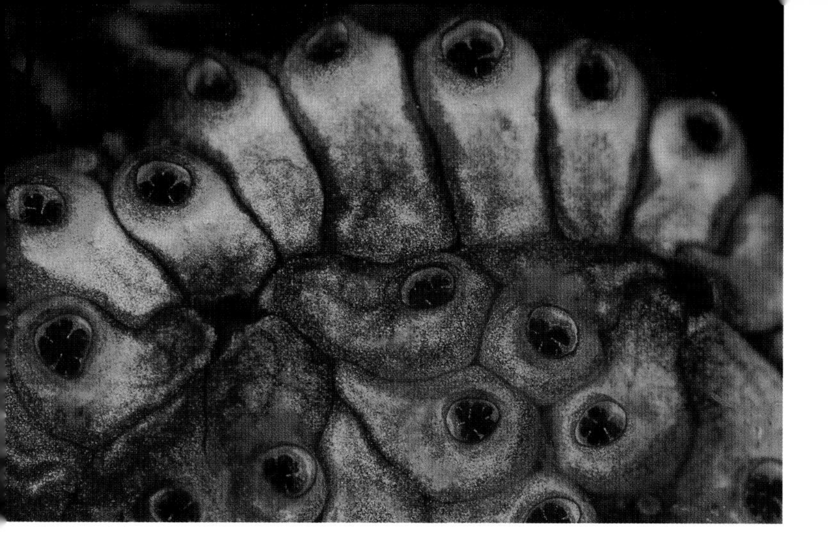

Botrylloides, probably *violaceus*
Phylum Chordata / Subphylum Tunicata / Class Ascidiacea

Compound tunicate, field of view about 6 mm. From floating dock, Monterey Harbor.

Like so many harbor animals, this usually brightly colored tunicate is a cosmopolitan "fouling" organism on boat hulls, floating docks, and even other organisms. This colony, for example, is growing on another fouling species of compound tunicate, *Diplosoma listerianum.*

The zooids of *Botrylloides* are small, about 1.5 mm long. As they are budded off, new members of the colony fit into meandering, serpentine systems of zooids arranged in parallel rows. These linear systems are easily distinguished from the rosette systems of most *Botryllus*, but the two genera are very closely related to each other and as a consequence have been classified inconsistently over the years. In contrast to *Botryllus* species, in which brooded embryos derive energy from their egg yolk, the eggs in this species of *Botrylloides* completely lack a yolk. The viviparous *Botrylloides* broods its embryos in a pouch, where they are bathed in maternal blood that provides the nutrients needed for larval development (Zaniolo et al. 1998).

Opposite: Adjacent *Botrylloides* and *Botryllus* colonies, field of view about 10 mm. From floating dock, Monterey Harbor.

As with *Botryllus*, adjacent colonies of *Botrylloides* will fuse when their fusibility genes match closely, though in nature fusible combinations occur less than 30 percent of the time (Mukai and Watanabe 1975). Remarkably, some *Botrylloides* colonies will even start fusing with adjacent *Botryllus* colonies. Immunologically, it's the invertebrate equivalent of grafting a human with a chimp's heart. Initially in these tunicates, there is a reciprocal blood vessel penetration of the two colonies followed by blood cell exchange. But then everything changes as rejection sets in, and one sees the signs of trouble ahead, as the zooids retreat from each other, with hemorrhaging and tissue death (Rinkevich, Lilker Levav, and Goren 1994). Our photo illustrates the only time we have found the two species next to each other; we could discern no apparent interaction between them.

Aeolidia papillosa

Phylum Mollusca / Class Gastropoda

Shag-rug nudibranchs, each about 35 mm long, and their egg strings. Piling, Municipal Wharf II, Monterey.

The anemone *Metridium senile* and its nudibranch attacker, field of view 8 cm. Piling, Municipal Wharf II, Monterey.

These sea slugs are specialists that prey on sea anemones. The "shag-rug" dorsal processes, called cerata, are actually outpouchings of the gut. On the wharf pilings, *Aeolidia* often occur in large numbers together, mating and laying their eggs and preying on the plumose anemone, *Metridium senile.* In compensation for their evolutionary loss of a protective shell, snails in this family, the eolid nudibranchs, ingest nematocysts (stinging cells) from their cnidarian prey and store these borrowed weapons in their cerata for later defensive use. In the nineteenth century, before this relationship was recognized, scientists believed that nematocysts, now recognized as an exclusive trait of the cnidaria, were also made by other animals, including nudibranchs.

All nudibranchs are hermaphrodites that cross-fertilize, usually by mutual copulation. During mating, they align so the sexual pores on the right anterior portion of their bodies are next to each other. Following the exchange of sperm and fertilization, they both lay egg strings, as these nudibranchs have done. *Aeolidia*'s eggs will hatch into swimming larvae and disperse in the plankton before settling and metamorphosing into juveniles.

Here, *Aeolidia* is gnawing on the foot of a small plumose anemone. Look closely and you will see that *Metridium*, in its defense, has extruded a tangle of threadlike gastric filaments (acontia) that are loaded with nematocysts. On contact with them, *Aeolidia* will secrete copious amounts of mucus that entraps the nematocysts and, at least for some types of nematocysts, markedly reduces their discharge (Mauch and Elliott 1997). But *Aeolidia* is playing with fire! Should the anemone's threads become entangled in its cerata, the nematocysts' stings may prove fatal. *Aeolidia* avoids large *Metridium* anemones, which may be too well defended. And when given a choice, the nudibranch takes other species of anemone—in particular, *Epiactis prolifera* and *Anthopleura elegantissima*—in preference to *Metridium.*

Note how closely *Aeolidia* resembles its prey. It soon takes on the anemone's color as pigments from this food source move into its cerata—a striking example of cryptic coloration. But even though *Aeolidia* stores *Metridium*'s nematocysts in its cerata and comes to resemble the anemone, some wrasses and flounders will find and feed on the nudibranch, though other fishes may reject it (Harris 1987).

Aeolidia does not have image-forming eyes, but it chemically detects anemones at a distance. Just as sharks respond to blood, these nudibranchs even show a marked preference for wounded prey. Clearly, these animals live in a world of chemical communication that far exceeds our own limited capabilities. Our own dependence on image-forming vision blinds us to the chemical ways in which many animals perceive and make sense of their worlds.

Phoronis vancouverensis
Phylum Phoronida

Right: Phoronid worms, field of view 17 mm. Piling, Municipal Wharf II, Monterey, 5 m deep.

Phoronis vancouverensis, which live in aggregations of chitinous tubes, are so inconspicuous that many divers have not knowingly seen them. The tiny worms feed with a horseshoe-shaped crown of ciliated tentacles, the lophophore, only about 3 mm across. Thus they are grouped with bryozoans and brachiopods as Lophophorates. The faint red blush at the base of some of the lophophores is due to the respiratory pigment hemoglobin in the animal's tissues. Phoronids are suspension feeders, capturing small planktonic particles from the water. Phoronids have an elongated, wormlike body, but their gut is U-shaped, doubling back on itself along the length of the body and ending with mouth and anus close together. As in bryozoans, the anus ends just outside the lophophore.

Phoronis is a simultaneous hermaphrodite, both male and female at once. Sperm packets released by one animal are caught in the lophophore of an adjacent one and taken into its body cavity. Here the packets burst, releasing masses of sperm that fertilize the eggs, probably just as they are released from the ovaries. The eggs, a few dozen to a thousand in number, are gathered and attached with mucus secretions to the base of the tentacles. In this photo, the paired, dense white masses within the inner whorls of the lophophores are brooded embryos. Thereafter, the embryos develop into distinctive planktonic larvae that precociously develop adult structures before they settle and metamorphose, often adjacent to existing adults. Almost immediately, the tiny phoronids form tubes from mucus secretions, thereby either enlarging the cluster of existing worms or establishing new clusters nearby (Zimmer 1997). The rapid colony formation and very tight clustering and dense aggregations of these phoronids, together with the relative scarcity of their planktonic larvae, suggest that they may replicate by fission as well as sexually. Although asexual reproduction of *Phoronis* occurs experimentally (Marsden 1957), it apparently has not been observed in the field.

3 The Kelp Forest

Viewed from within, the kelp forest is a magical place. Slipping weightless beneath the dense, tangled canopy, you glide into a natural cathedral. Entwined stipes of kelp rise in tall columns; rays of light stream down through the water. As you advance, schools of blue rockfish break ranks. The physical effort of propelling yourself and your equipment, the pervasive cold, the rushes of adrenaline—the experience is demanding, addicting, and rewarding, if at times intimidating.

Macrocystis pyrifera—giant kelp—rapidly growing and enormously productive, makes up the forest. It dampens surge (the underwater counterpart of wind gusts) and so provides relative shelter, making a home for the sea otter and a nursery ground for fishes. Kelp blades provide a leaflike substratum for countless organisms that in turn become food for legions of browsers. The kelp itself provides a seemingly inexhaustible pasture for herbivores. When kelp fronds in the forest senesce, or when they decompose on the beach, they form detritus, organic particles that become food for suspension-feeding animals; they also release tremendous amounts of accumulated chemical nutrients into the water.

Kelp clings to a foundation of rock by means of its holdfast. The rock is buried beneath a luxuriant growth of algae and attached solitary and colonial invertebrates that are busily capturing food and releasing gametes and larvae. Mobile herbivores and carnivores creep over the sessile organisms. The diversity of life in a kelp forest rivals that of the land's most exuberant habitats. As Darwin (1839) wrote while sailing off the coast of Chile, "I can only compare these great aquatic forests of the southern hemisphere with the terrestrial ones in the intertropical regions. Yet if the latter should be destroyed in any country, I do not believe nearly so many species of animals would perish, as, under similar circumstances, would happen with the kelp." Without its kelp forest, to be sure, Monterey Bay would be profoundly impoverished.

The "cathedral."

Macrocystis pyrifera

Phylum Heterokontophyta / Class Phaeophyceae

Opposite: Macrocystis frond, about 30 cm, seen from 5 m deep. San Clemente Island, California.

Kelps are large brown algae, and *Macrocystis* is gigantic: these seaweeds can grow from depths of over 25 m to reach and spread along the surface in a dense canopy. During upwelling, they can grow at a phenomenal rate, up to 50 cm a day, as fast as tropical bamboo. Gas-filled floats support the growing fronds of the anchored *Macrocystis* as the kelp proliferates in a race toward the sunlight above.

Like all plants and all seaweeds, brown algae need chlorophyll as well as sunlight for photosynthesis. Essentially, in photosynthesis the energy of light is converted to the energy of carbohydrate molecules. The green color of kelp's chlorophyll is masked by the golden brown carotenoid pigment fucoxanthin, an accessory photosynthetic pigment. Fucoxanthin and other such accessory pigments act as solar antennae: they capture the energy of light wavelengths other than those trapped by chlorophyll and funnel it to the chlorophyll, thus enabling photosynthesis to proceed. Seawater rapidly absorbs the red end of the solar spectrum, but sunlight's green wavelengths penetrate coastal seawater deeply, and these are precisely the wavelengths that fucoxanthin absorbs most efficiently.

A freshly cut section through the stemlike stipe of *Macrocystis* exudes quantities of slimy secretions from "sieve tubes," structures that conduct the kelp's products of photosynthesis and minerals in a manner analogous to the movements of sap in vascular plants. Organic products move through the sieve tubes in the kelp's stipe much faster than by diffusion alone—in fact, at a rate comparable to that in vascular plants (Parker 1971). Among the nonvascular plants, only the large kelps like *Macrocystis* have sieve tubes.

At a certain depth below the surface known as the compensation depth, a seaweed's rate of respiration and its rate of photosynthesis equal each other. Because photosynthesis is dependent on sunlight, this depth varies with the density of the kelp canopy and the clarity of the water as well as by season and hour. In large kelps, maximum photosynthesis occurs in mature blades that are bathed in abundant light near the surface. Here the products of photosynthesis exceed the energy needs of the alga for respiration; excess photosynthetic products therefore move down through the stipe's sieve tubes to the dimly lit basal portions of the alga below the compensation point, where these products are needed. Above the compensation point, excess organic products also move upward and out to those frond tips that exceed their local energy resources during periods of especially rapid growth.

Macrocystis apical scimitar blade, 18 cm. Monterey Bay, 3 m deep.

This specialized, expanded blade at the tip of each *Macrocystis* frond contains at its base the persistent growth zone, or meristem, of the seaweed. The blade forms a protective shield over the small area where the leaflike blades and floats develop and split apart as the kelp grows toward the surface of the sea. The white spots on the new blades in this photo are newly settled colonies of the encrusting bryozoan *Membranipora membranacea.*

Opposite: Macrocystis holdfast, field of view about 1.4 m. Carmel Bay, 17 m deep.

Giant kelp is securely anchored to rock with a cone-shaped holdfast composed of very tough tendrils called haptera, thus withstanding the drag of ocean surge and the lift of its floats. Rather than being nourished by feeding roots as vascular plants are, giant kelp absorbs nutrients directly from the water, principally through its blades. It does not grow in quiet water but depends on water flow to bring the nutrients needed for rapid growth. The holdfast shown here is surrounded by encrusting coralline red algae, sponges, and compound tunicates.

On average a giant kelp lives for six to eight years, during which time a community of tiny creatures will exploit the safety of refuges among the haptera. The most abundant of these are molluscs, polychaetes, brittle stars, and crustaceans such as isopods and amphipods (Vasquez 1993). During the *Beagle*'s cruise through Tierra del Fuego, Darwin (1839) was struck by the *Macrocystis* holdfast community: "On shaking the great entangled roots, a pile of small fish, shells, cuttle-fish, crabs of all orders, sea-eggs, starfish, beautiful Holuthuriae (some taking the external form of the nudibranch molluscs), Planariae, and crawling nereidous animals of a multitude of forms, all fall out together." Some of these creatures are herbivores. By burrowing into haptera and feeding on them, they eventually hollow out and weaken the holdfast. During storms these holdfasts fail and the huge bodies of kelp are ripped from the bottom, pulling others with them as they entangle themselves in the forest canopy and are driven toward shore.

Giant kelp is the diploid sporophyte of the alga's life cycle (Fig. 7). In specialized blades clustered immediately above the holdfast the kelp produces huge numbers of haploid male and female spores, with a single blade producing up to 500,000 spores an hour. The spores, each propelled by two flagella, settle and

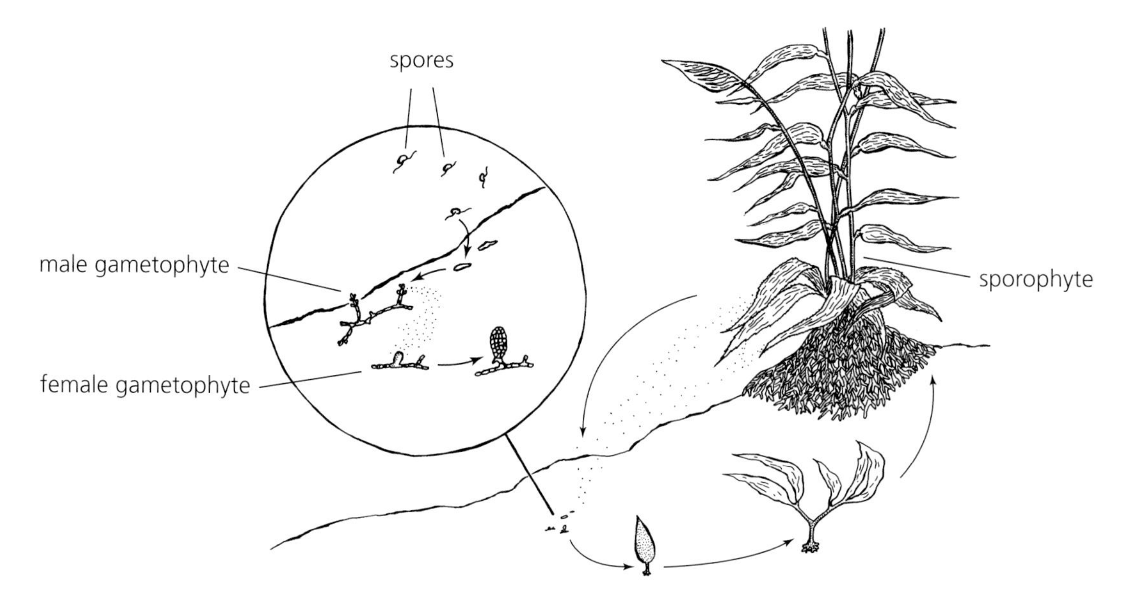

Fig. 7. *Macrocystis pyrifera* life cycle

germinate into the short-lived and minuscule haploid male or female gametophyte stage of *Macrocystis*'s life cycle. Spores often settle within a few meters of their release and in a density of up to 200 spores/cm^2 (more than 1,200 per square inch!). Pulses of spores released by different *Macrocystis* sporophytes also seem to be synchronized by environmental factors such as storms, promoting even denser settlement (Reed et al. 1997). This proximity of gametophytes to each other enhances the chances of eventual fertilization among their gametes.

All that makes sense, but how does one explain dense new stands of *Macrocystis* sporophytes several kilometers from the nearest living fertile adult? It has long been argued that floating rafts of the torn-up plants must have released their spores near such new sites. Now, however, it is clear that *Macrocystis* spores may disperse much more widely than previously thought. They can photosynthesize the energy they need (much as larvae gain energy by feeding) and can respire and swim for as long as three days, even germinating while afloat (Reed, Amsler, and Ebeling 1992). That the resulting gametophytes grow densely enough to cross-fertilize is a testimonial to the vast numbers of spores that their parents synchronously spewed out, and perhaps to currents that herded spores in the same direction.

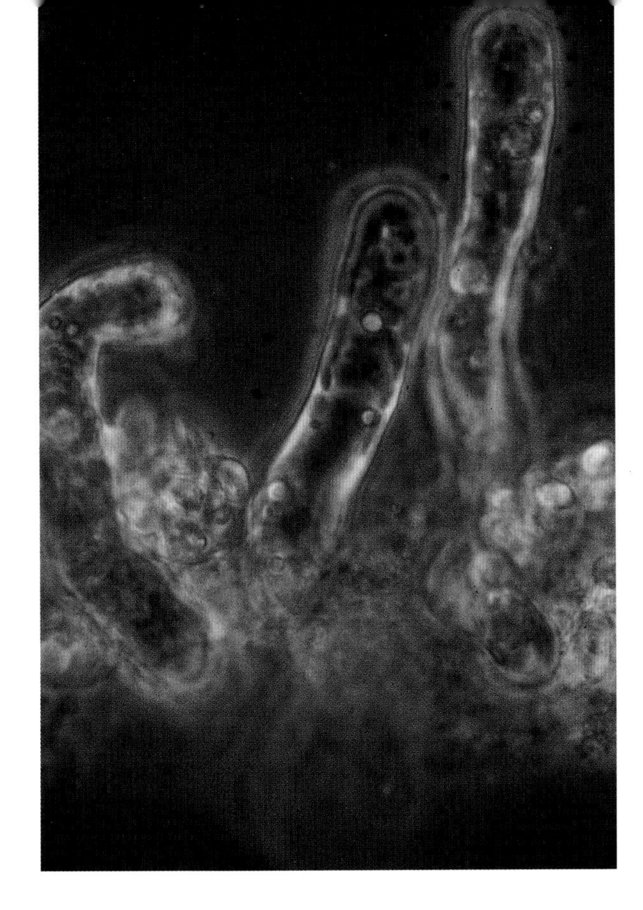

Macrocystis gametophytes from culture, sex undetermined, about 50 microns long. (Photo courtesy of Moss Landing Marine Laboratories)

The contrast in size between these little gametophyte filaments, too small to see without magnification, and their giant sporophyte progeny staggers the imagination. The future giant sporophyte grows directly from the microscopic fertilized female gametophyte.

For the sperm released by male gametophytes, the search for a partner is not completely blind. Many brown algae, including *Macrocystis* (and *Fucus*), have pheromones in their ova that act as sperm attractants. One of them, lamoxirene, isolated from the ova of *Macrocystis* and several other large kelps, also causes the explosive release of sperm from male reproductive structures. So two steps are involved in the chemical coordination of fertilization: the pheromone first causes sperm release and then attracts sperm to the eggs.

Algal sperm may be remarkably sensitive to these pheromones, responding to as little as a single molecule; and that is just as well, because a single *Macrocystis* egg secretes only 10^{-12}–10^{-14} moles of the hormone (Maier and Müller 1986). To put it another way, five million *Macrocystis* eggs yield only 2.9 micrograms of lamoxirene (Lobban and Harrison 1994). Additionally, lamoxirene may exert its effect over a distance of about 0.5 mm; this observation correlates well with the experimental finding that for new sporophytes to become established, populations of gametophytes must have a density of at least $1/mm^2$ (Reed 1990). Although the pheromones throughout this group of related brown algae are apparently identical, hybridization between species occurs rarely in the field.

Ferns, which are vascular plants, have a life cycle strikingly similar to that of *Macrocystis.* Here, too a generation of sporophytes alternates with a generation of gametophytes. In this case, the enchanting leafy fronds are the sporophytes; they meiotically produce spores in the rows of little dots, or sori, you have seen on the undersides of fern fronds. Winds disperse the spores, producing the inconspicuous gametophytes that bear both male and female gametes. Fertilization, of course, gives rise to more sporophytes.

Throughout this book, you will encounter many algae and invertebrate animals that thrive as they work their way through such complex life cycles of alternating sexual and asexual forms.

Wrack. Carmel Beach.

After a winter storm, great piles of seaweeds have been cast up on Carmel Beach. Note the tangled mass of giant kelps and their holdfasts, together with several bull kelps. Soon herbivores, including fly larvae (maggots) and amphipods, and decomposing bacteria and fungi will go to work and, within a few days, reduce this algal mass to a heap of detritus and dissolved organic matter. Thrust your arm into a big pile of stranded kelp and you will be startled by the high temperature inside; digestion of the pile proceeds in a veritable oven heated by bacterial metabolism. If you detect the odor of rotten eggs it's because of anaerobic decomposition, which releases hydrogen sulfide gas. Subsequent waves and tides will return huge amounts of this organic material to the sea as the piles break up and slide awash. Suspension-feeding creatures and other animals soon devour these vital nutrients. The edge of the sea during these recyclings is one of the richest places on earth.

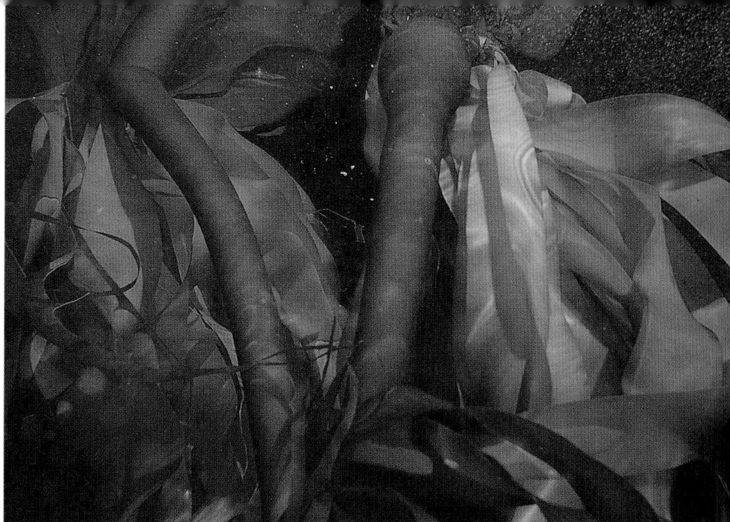

Nereocystis luetkeana
Phylum Heterokontophyta / Class Phaeophyceae

Bull kelp, field of view 50 cm. Carmel Bay, 2 m deep.

In shallow water, glorious sunlight bathes these two bull kelps, which arise in dim light from a depth of 20 m. One float bobs just below the sea's surface; the other, concealed here by reflections, is thrust through into the air above. When viewed from shore, these bobbing floats often are mistaken for the heads of sea otters. *Nereocystis*, in contrast to the giant kelp, *Macrocystis*, is an annual. In the summer it continues its rapid growth, often at more than 25 cm per day, rather than accumulating storage material that it will not need (Kain and Norton 1987). Growth is fastest during daylight hours, but continues in the dark as well. As *Nereocystis* approaches the surface, its growth slows and spore production begins.

We find *Nereocystis* in exposed, turbulent areas, often at the edges of giant kelp beds. Each thallus consists of a single long, slender, hoselike stipe that arises from a small holdfast, widening until it culminates at the sea surface in an elongate, bulbous, gas-filled float from which dangle clusters of smooth blades, as seen here. Not surprisingly, the float contains oxygen (from the plant's photosynthesis) and nitrogen, but it also contains 4 percent carbon monoxide (CO) by volume. S. C. Langdon (1917) confirmed the presence of CO not only by analytical methods but also by performing an autopsy on a guinea pig that succumbed after inhaling the gas—an experimental methodology that might not be acceptable today. The function (if any) of CO isn't clear, though herbivorous animals with respiratory pigments that could be poisoned by CO seem unlikely to burrow into the float.

In late summer or autumn, spores form by meiosis on *Nereocystis* blades in patches called sori. The sori detach, usually at dawn or shortly thereafter, and drift down to the seabed, releasing many of their photosynthetic spores during their descent and thus ensuring a maximum opportunity for dispersal by currents. Spores continue to be released for several hours after the sori come to rest; these spores will germinate near the parent and close to each other (Amsler and Neushul 1989). Not all *Nereocystis* sori reach the bottom; blue rockfish consume many of them during their drifting descent.

As with *Macrocystis*, so also with *Nereocystis*, the proximity of gametophytes to each other enhances prospects of fertilization and new sporophyte production. It also helps that the eggs of *Nereocystis*, like those of several brown algae we discuss (*Macrocystis* and *Fucus*), produce the powerful pheromone lamoxirene, which, even in an extremely small amount, causes an explosive release of sperm from male gametophytes and then acts as a sperm attractant (Maier and Müller 1986).

During winter storms, *Nereocystis* washes up on beaches, tangled with *Macrocystis* in large piles of wrack that can be closely inspected. Often the stipes of last season's *Nereocystis* are covered with algae such as the red *Porphyra nereocystis*, which we discuss next.

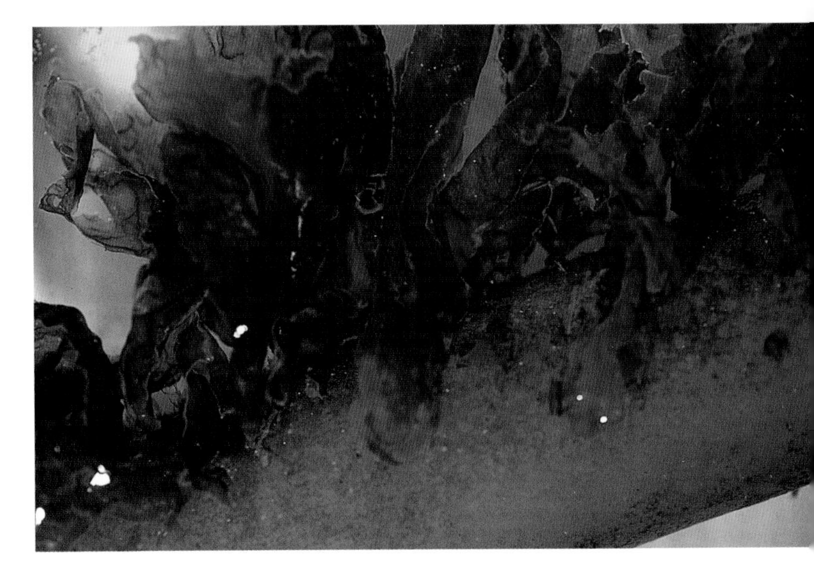

Porphyra nereocystis

Phylum Rhodophyta / Class Bangiophyceae

Red algal gametophytes, to 16 cm long, on stipe of *Nereocystis luetkeana*. From Carmel Bay.

Porphyra is a specialized epiphytic red alga with distinctive heteromorphic phases in its life cycle (Fig. 8). The gametophytic blades of this species grow only on the stipe of the bull kelp *Nereocystis luetkeana*, as in this photo. The sporophyte phase, so-called *Conchocelis*, lies hidden, growing on empty mollusc shells on the seafloor. These strikingly different phases, once considered two separate algal species, somehow coordinate their own reproductive activities with those of their kelp host. In the spring, "*Conchocelis*" releases diploid "conchospores" in strands of mucus that attach to the stipe of a young bull kelp just beneath the float. They are then carried almost to the sea's surface by the kelp's growth; there they meiotically produce the rather luxuriant, haploid gametophytic blades shown here. Eggs and sperm develop on the same blade, and, following fertilization, the red alga's third phase, diploid carposporophytes, grow from the zygotes and release their spores. These carpospores drift down to the seabed, somehow find an empty mollusc shell, and become the shell-boring, filamentous "*Conchocelis*" sporophyte phase (comparable to the tar spot "*Petrocelis*" phase of *Mastocarpus,* but far smaller). We would love to know how these spores find a shell. Can it be solely a matter of chance? As far as we know, this interesting puzzle has not yet been investigated.

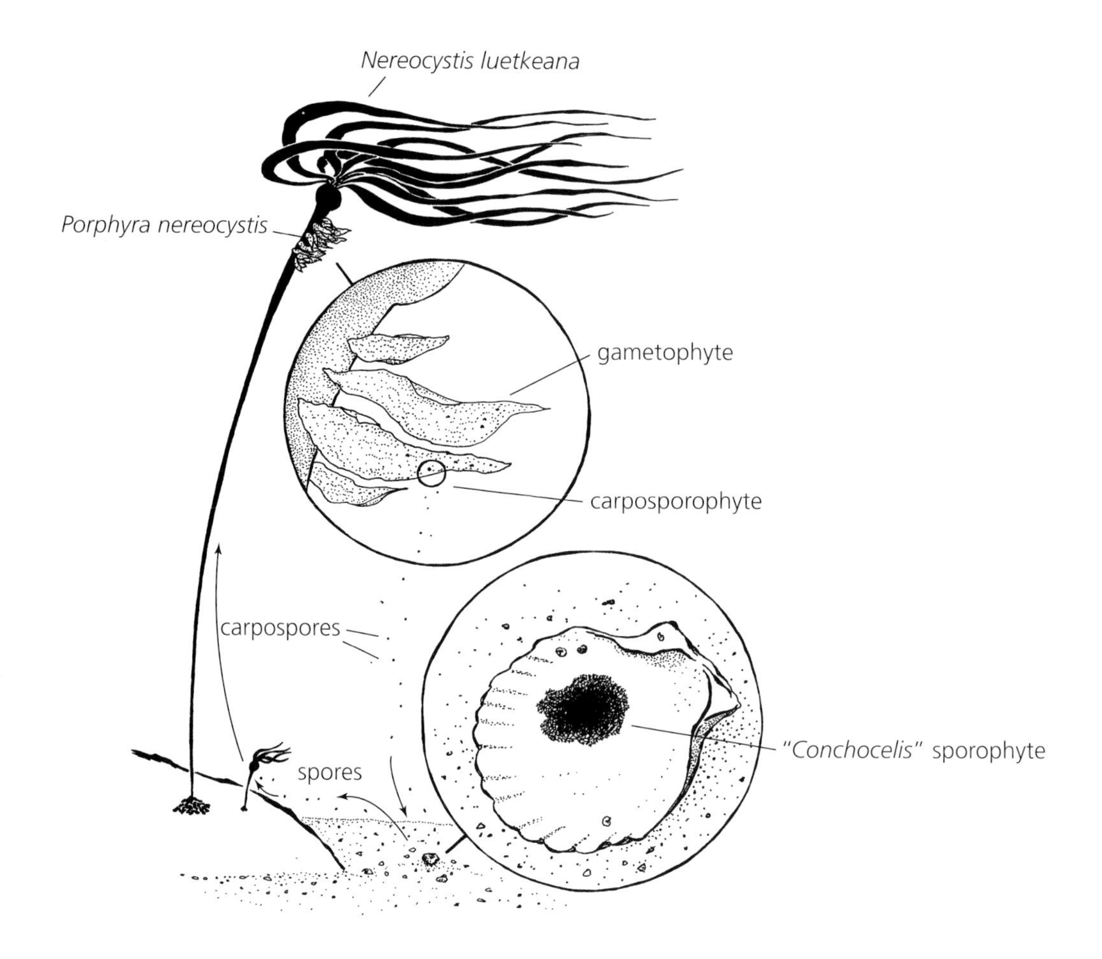

Fig. 8. *Porphyra nereocystis* life cycle

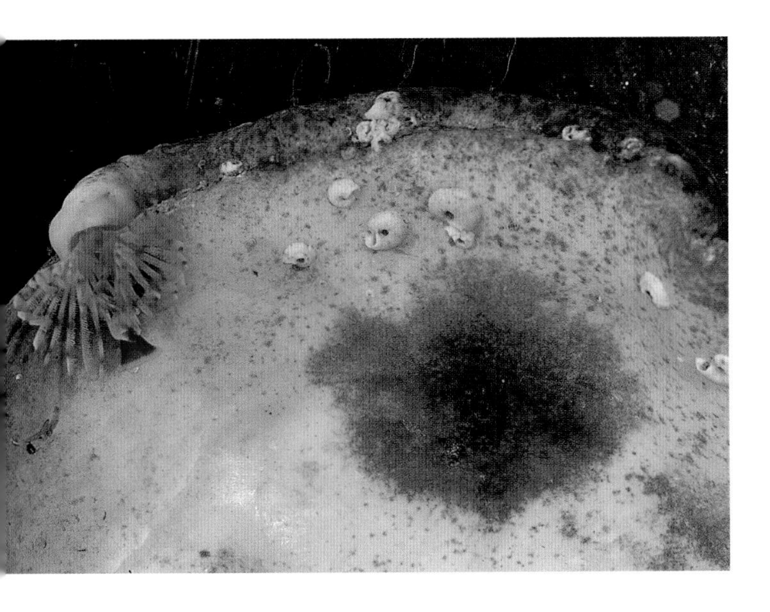

"Conchocelis rosea," 10 mm across, on a scallop shell. From Monterey Bay.

This pink patch, growing on a scallop shell, is "*Conchocelis*," actually the sporophyte phase of *Porphyra*, apparently the perennial stage in the alga's life cycle. It consists of tiny, branching filaments that bore into its shell host. The alga in the photo is immature and does not yet bear conchospores; once it has matured, in the spring it will release its conchospores and they will find and attach near the float on young, short *Nereocystis* stipes. The kelp's rapid growth will then carry that preferred site up and away to the distant sea surface. The sporophytes accomplish this feat by somehow "measuring" the daylight hours: a prolonged period of short winter days followed by lengthening spring days stimulates conchospore release in timing with the bull kelp's own life cycle (Dickson and Waaland 1985).

Note the plume of the polychaete worm *Serpula vermicularis* in the upper left; the worm's calcareous tube runs along the scallop shell's edge. The little coiled tubes of several spirorbid worms also are visible, most of them empty.

In Japan, another species of *Porphyra* spends its foliose gametophytic phase on intertidal rocks. It is extensively collected as nori, a food high in protein, with more vitamin C by weight than an orange, and high in iodine and other trace elements. We usually encounter nori as sushi wrappers. The widespread commercial production of nori in Japan became possible only when an English botanist, Kathleen Drew Baker, showed that the genus "*Conchocelis*" is not a separate species of alga but actually a phase in *Porphyra*'s life cycle (Drew 1949). In Japanese bays today, fishermen seed nets with "*Conchocelis*" filaments grown in oyster shells and harvest the resulting gametophytic blades as nori. Kathleen Drew Baker is commemorated with a monument at Hiroshima Bay.

Red algal life cycles seem complicated, yet complexity does not belie their extraordinary evolutionary success. Tiny red algal fossils were reported recently from sediments laid down 570 million years ago in China (Xiao, Zhang, and Knoll 1998). The algal cell clusters are interpreted as showing carpospore formation in thalli indistinguishable from contemporary *Porphyra!* The soft tissues are preserved in exquisite cellular detail, as they were quickly replaced by calcium phosphate during petrifaction.

Enhydra lutris
Phylum Chordata / Subphylum Vertebrata / Class Mammalia

Male southern sea otter, estimated 25 kg. Monterey Bay.

This grizzly-headed sea otter has assumed a characteristic resting posture. Sea otters often float on their backs this way, entangled with kelp, legs held aloft and head raised. In this position they are less exposed to the pervasive cold water, and the kelp ensures that they don't drift away. The heavy whiskers on the sides of its snout are important sense organs. Under voluntary control, they can be extended forward to feel—presumably a useful technique during dives when visibility is poor. A sea otter's nostrils, at the front of its snout, close during dives. This otter has recently surfaced: his fur appears wet and shows points where guard hairs stick together; the underfur, however, is quite dry. A sea otter's fur is much denser than that of other marine mammals. This luxuriousness is essential to reduce heat loss, as otters lack underlying blubber. They must groom several hours a day, rubbing themselves with their forepaws to keep their fur fluffy with heat-insulating air, rolling and somersaulting to cleanse themselves, all to keep their pelage in prime condition. Also, a high metabolic rate helps maintain their body temperature. An oil-soiled sea otter rapidly dies of hypothermia, and just three days without food is enough to cause death by starvation. To maintain this internal furnace, a sea otter consumes food averaging 25 percent of its body weight daily, equivalent to 2.5 tons of invertebrate biomass annually.

Note the large webbed hindpaws. Sea otters at the surface swim on their backs using strokes of these flippers and vertical movements of the flattened tail. One mother was observed carrying her pup and swimming at 2.4 km/hr (Kenyon 1975). For such semiaquatic animals, however, surface swimming is energetically very expensive: an otter swimming at the surface uses five times as much energy as a sea lion of similar size swimming underwater (Alexander 1999). On dives, sea otters can briefly swim at twice that speed as they vigorously undulate their body in addition to paddling.

Both the forepaws and hindpaws have five digits.

The forepaws have no opposable thumb, and the "fingers" cannot be used individually because they are held together by webbing; in fact, they scarcely flex. Because of these features, together with the large palmar pads so obvious in the photo, otters are unable to grasp small objects with a forepaw. However, their wrists are so flexible they can easily grasp objects between the paw and the forearm; large objects are readily carried with two paws or under an arm. A skilled hand surgeon, L. D. Howard, whose dissections of sea otter limbs demonstrated many of these facts, pointed out that their forefoot was designed mainly for running (1973). It is not surprising, therefore, that of the marine mammals, they are the most recently evolved for life in the ocean. Nevertheless, sea otters dexterously use their forelimbs for eating, grooming, and caring for their young. They are also among the few animals that habitually use tools. On the sea bottom, they will use a rock to retrieve abalone by smashing its shell; at the surface they often balance a rock on their chest as an anvil upon which they break open molluscs and crabs. A sharp cracking sound from a nearby kelp bed often signals an otter feeding offshore.

Above right: "Jaws," 12 cm.

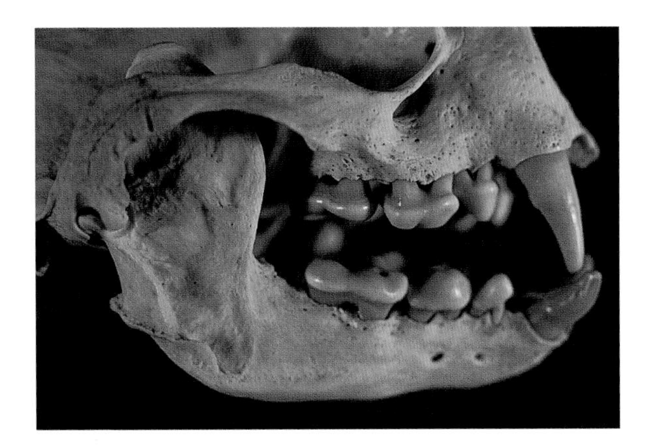

This is a sea otter's cranium—evidently an adult's, as most of the sutures are obliterated, and probably a female's, for it is relatively small. In otters, the two halves of the lower jaw do not fuse but form a highly mobile joint. The lower jaw articulates snugly with the base of the skull. These features help otters use their teeth to grasp, crush, and hang on to hard objects. Sea otters use their spade-shaped lower incisors (some of which are missing here) to scoop clams or sea urchins from their shells. Their molars are flattened and rounded for crushing hard-shelled invertebrates, their favorite prey. In contrast, the molars of many other carnivores have sharp cusps (look in the mouth of your dog or cat), useful for cutting or shearing flesh. The otter's canines are round and blunt; they are used to pry open bivalves (Kenyon 1975).

The teeth of this otter have a definite purple tinge—most noticeable when compared to truly white teeth—owing to pigments derived from eating purple urchins (or perhaps from the milk of its mother who ate urchins) during the teeth's formative stage. Although this skull is only lightly stained, some otters' bones and teeth are deeply colored by these pigments (Judson Vandevere, pers. comm.).

Most sea otters specialize in only one to three prey species, such as kelp crabs, rock crabs, turban snails, red abalone, urchins, or mussels. When she pups, a female sea otter often changes her foraging habits to focus on easily captured prey, possibly to teach her youngster these techniques (Lyons 1986).

Sebastes mystinus
Phylum Chordata / Subphylum Vertebrata / Class Osteichthyes

Blue rockfish, each about 30 cm. Carmel Bay, 5 m deep.

These blues were a dramatic sight as they hung almost motionless, quietly suspended in the kelp forest, looking very relaxed. But when we kicked our way closer, the school rapidly broke ranks. How did they do this? They were utilizing organs that mammals don't even have: a swim bladder and a lateral line.

Rockfishes, and indeed most bony fishes, adjust their buoyancy with swim bladders that they inflate with gases, mostly oxygen, which is metabolically released from the blood pigment hemoglobin (Baines 1979). These midwater fishes remain neutrally buoyant at different foraging depths by altering their swim bladder's gas pressure, just as divers do when they add or release air from their buoyancy compensator.

How do rockfish maintain their relative positions in their school? To some extent they do it visually, of course, but more importantly they sense their surroundings via their lateral line organ, a neural line along each side of the head and body that detects pressure changes, even the pressure waves of sounds. The main line of the system consists of a branch of the vagus (tenth cranial) nerve; the nerve's sensory cells have minute hairs enclosed in gelatinous sacs lying in shallow canals in the skin. Any water movement displaces the sacs and so stimulates these sensory cells. The system permits the fish not only to school but also to detect predators, even when visibility is poor. The astonishing whole-school maneuvers that many fish perform probably reflect exploitation of this sensory organ, the way purely visual cues may control flocking behavior in birds.

Rockfishes are members of the scorpionfish family. Many scorpionfish have highly venomous spines, but those of blue rockfish are only mildly so. Blues are by far the most abundant of the many rockfish species in Carmel Bay's inshore waters. In the spring and summer they feed principally on large gelatinous planktonic organisms such as medusae, salps, pelagic snails, and comb jellies. They also take crustaceans and occasionally fish such as anchovy, but rarely organisms smaller than 5 mm. In the fall, in the absence of their preferred prey, blues grab the reproductive parts (sori) of *Nereocystis* sporophytes when these detach and drift down toward the seabottom, and they also ingest the delicate red algae *Porphyra nereocystis* and *Smithora naiadum* (Hobson and Chess 1988; Hobson, Chess, and Howard 1996). A blue's rather small mouth and long gill rakers make it an effective predator on plankton; its very long gut and unusually large and numerous absorptive villi in the stomach and intestine—features also of herbivorous fishes—may allow it to digest algae efficiently. This diet overlaps little with that of other rockfishes, and blues regularly coexist with both black and olive rockfishes (Hallacher and Roberts 1985).

All rockfishes copulate. Females can store sperm from copulation until ovulation. The females brood the young for a short time after the eggs have hatched within the maternal body. The larvae then spend several months at sea in the plankton before returning to the kelp beds as juveniles.

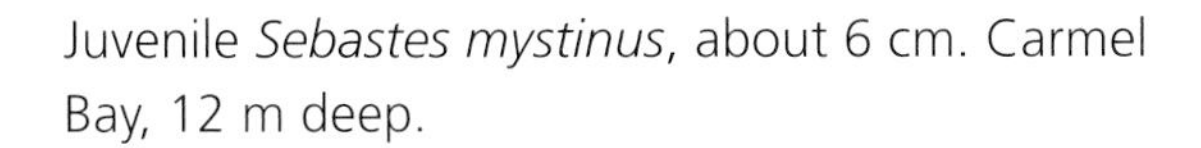

Juvenile *Sebastes mystinus*, about 6 cm. Carmel Bay, 12 m deep.

Since adult blues are so numerous, during favorable years there are great numbers of juveniles. But they are not blue: they are pinkish red, and at depth they appear gray. During the day we have often seen juveniles, but not adults, aggregated close to the bottom, which may promise refuge from predators.

The rockfish fishery is in a decline, and we see these fish less frequently than when we started diving central California waters in 1980. There is a growing consensus that rockfish of all kinds have been overfished and that conservation measures are now needed to restore their populations. Young-of-the-year blues are most numerous during the spring and summer upwelling season, at which time they make up a large part of the diet of other rockfish species, lingcod, and cabezon. So a sustained decline in the number of blue rockfish could spell trouble in turn for many of the piscivorous species in the kelp forest, including other rockfishes.

Oxyjulis californica
Phylum Chordata / Subphylum Vertebrata / Class Osteichthyes

Señorita fish, length 20 cm. Monterey Bay, 15 m deep.

These little wrasses, adults of which are usually associated with the kelp forest, pick at their food with their small mouths and buckteeth. Señoritas are famous for their habit of cleaning the bodies of other fish. Typically, a host fish (say, a kelpfish) stops swimming, erects its fins, and drifts passively, head down. The señorita then removes the fish's parasitic copepods and isopods. For the bulk of their diet, however, señoritas prey on invertebrates living on kelp blades. They efficiently pick off erect, attached organisms such as hydroids and catch small crustaceans such as amphipods and isopods in the canopy and in midwater. Encrusting, flat colonies such as the bryozoan *Membranipora membranacea*, a favored prey item, are more problematic. Instead of scraping the animals off the encrusted kelp blades, something they are unable to do, the señoritas simply nibble off the blade's edge. In this way they inflict considerable damage at the edges of kelp bed canopies, where *Membranipora* is abundant; *Oxyjulis* has even been known to wipe out small kelp bed canopies (Bernstein and Jung 1979).

Señoritas are diurnal fish. At dusk they roll on their sides and bury themselves by swimming head-

first into the loose sand, only to emerge again at sunrise. Frightened fish may also seek escape by similarly diving into the sand. Although we have few nocturnal predatory fish compared to the tropics (the torpedo ray comes to mind as one exception), the señorita may retain the burying habit of its family as genetic baggage from the past rather than as a trait molded by the selective pressure of nocturnal predation here and now (Bray and Ebeling 1975).

These fish (and probably many others) "know" where they are. When, in one experiment, señoritas were tagged and translocated to alternative kelp bed sites, 80 percent returned to their home kelp beds (Hartney 1996).

Señoritas spawn great numbers of eggs that are fertilized externally. The larvae develop pelagically for about eight weeks until the few that have survived this especially vulnerable phase of the life cycle return to the kelp forest as juveniles. They do not change sex, though many other wrasses do.

A Kelp Blade Community

Some "fouling animals" that live on giant kelp blades, field of view 16 mm.

Here we see a rich assemblage of little invertebrates forming a community on a blade of giant kelp in the surface canopy (Fig. 9). Kelp blades provide a relatively huge, if transient, stage for a life-and-death drama of tiny actors. In spring and summer the stage is continually replenished by the kelp's growth, but later it is reduced and even wholly removed by the loss of canopy in autumn die-back and winter storms. This cycle places many constraints on the community. Since kelp blades live for an average of three months, the sessile animals that colonize them must conform to an exacting timetable of rapid growth, early sex, and brief life.

It has been rewarding for us to scoop up kelp blades and scrutinize them in our laboratory. This photograph was taken in May during the upwelling season, when the ocean is cloudy with plankton.

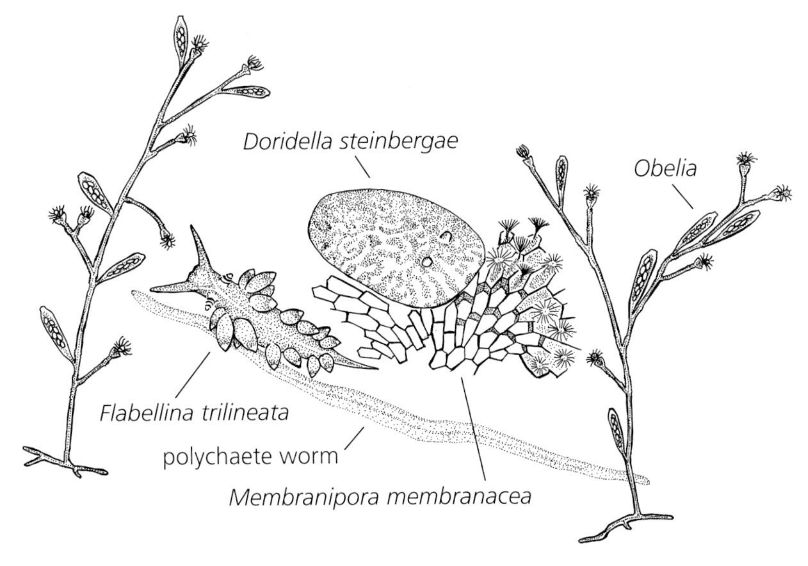

Fig. 9. A kelp blade community

These animals all arrived as larvae out of the planktonic soup. Look for tiny grids, fragmented colonies of the bryozoan *Membranipora membranacea;* adjacent to them is one of their nudibranch predators, *Doridella steinbergae,* which has been devouring them.

On the blade's surface are strands of male and female colonies of the hydroid *Obelia.* The colonies' reproductive polyps are full of medusa buds, resembling tiny beads on a string, that await release as little jellyfish. With magnification, it's exciting to observe the little medusae wriggle free from their parental polyps and rapidly pulsate away. In this view, the normally erect *Obelia* polyps are pressed under a glass cover-slip, posing for their photograph. (See Figure 10 for *Obelia*'s typically hydrozoan life cycle.) The nudibranch *Flabellina trilineata,* a predator on hydroids, lurks nearby. Soon, many such predatory nudibranchs will appear on the scene and wipe out this generation of *Obelia.* Finally, a wandering polychaete, which probably also settled as a larva, is slithering its way across the field of view.

These kelp blade communities did not escape Darwin's penetrating attention (1839): "The number of living creatures of all orders, whose existence intimately depends on kelp, is wonderful. A great volume might be written, describing the inhabitants of one of these beds of sea-weed. Almost every leaf, excepting those that float on the surface, is so thickly incrusted with corallines, as to be of a white colour. We find exquisitely-delicate structures, some inhabited by simple hydra-like polypi, others by more organized kinds, and beautiful compound Ascidiae. On the flat surfaces of the leaves various patelliform shells, Trochi, uncovered molluscs, and some bivalves are attached. Innumerable crustacea frequent every part of the plant."

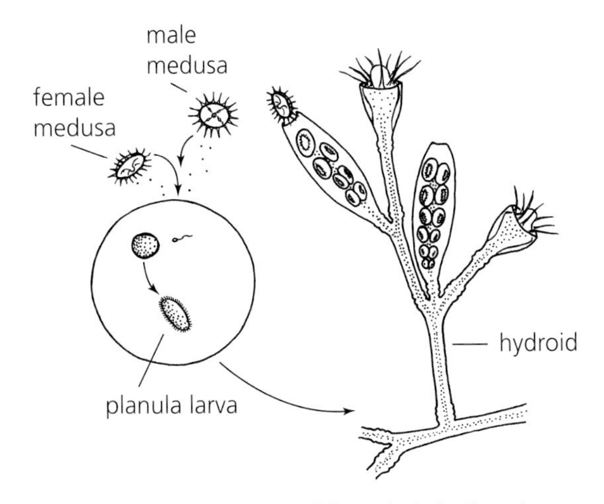

Fig. 10. Hydrozoan life cycle (*Obelia* sp.)

Membranipora membranacea

Phylum Bryozoa / Class Gymnolaemata

Opposite: Bryozoan colonies on frond of giant kelp, about 75 cm. Carmel Bay, 5 m deep.

This kelp frond is heavily encrusted with the bryozoan *Membranipora.* Such infestations are common in the kelp forest. *Membranipora* colonies consist of tiny zooids, living in calcareous "houses" less than 1 mm long. Replication of the zooids by budding leads to the colony's rapid spread, an important process since there is very little time for the bryozoan to go from larval settlement and metamorphosis to sexual spawning before the kelp frond it is living on disintegrates. Although *Membranipora* can live on other surfaces, given a choice, the larvae prefer settlement on brown algae. The colonies are heavily preyed upon by several species of nudibranch and by señorita fish, *Oxyjulis californica;* but as we shall see, the bryozoan has a clever defense.

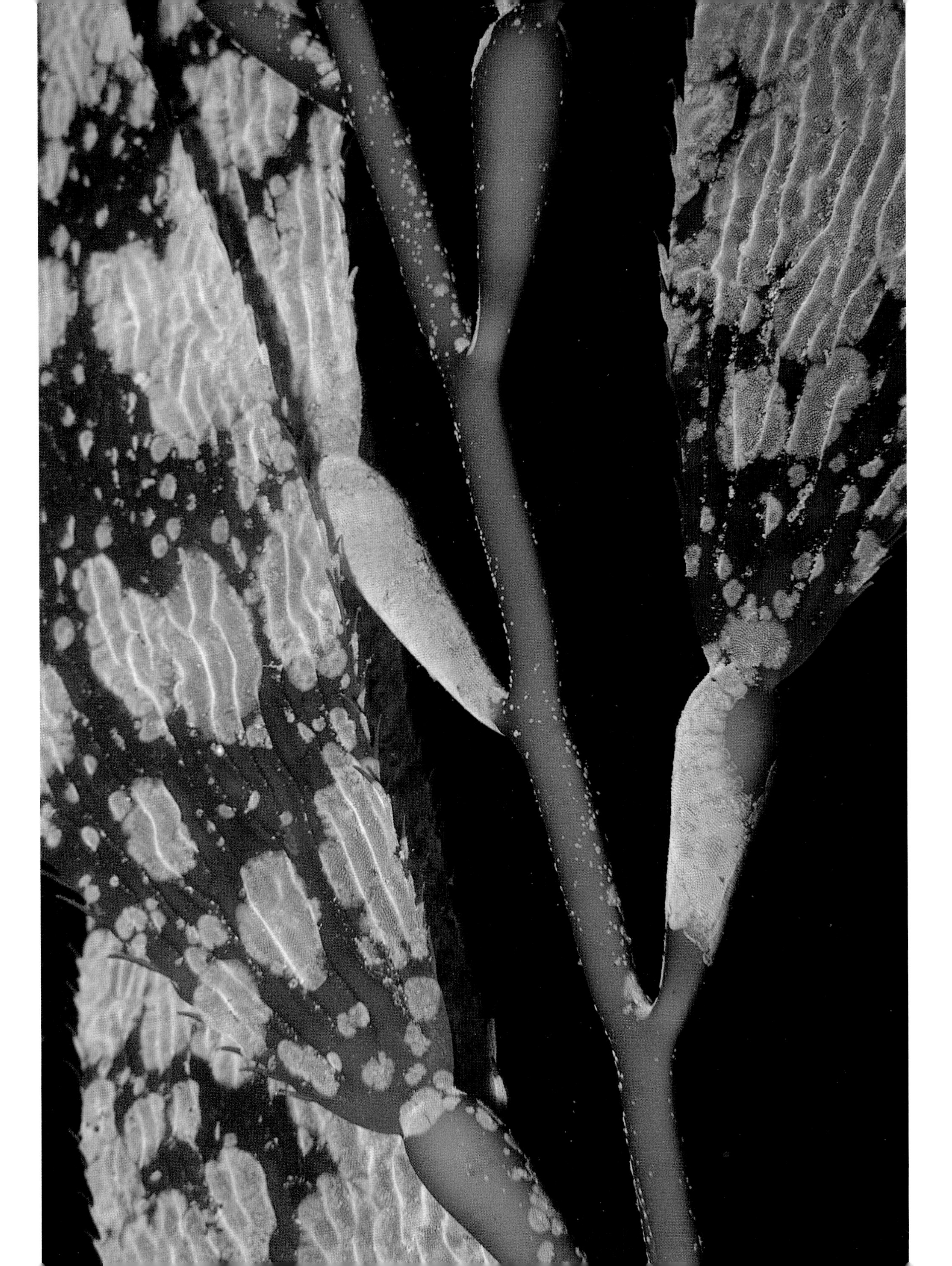

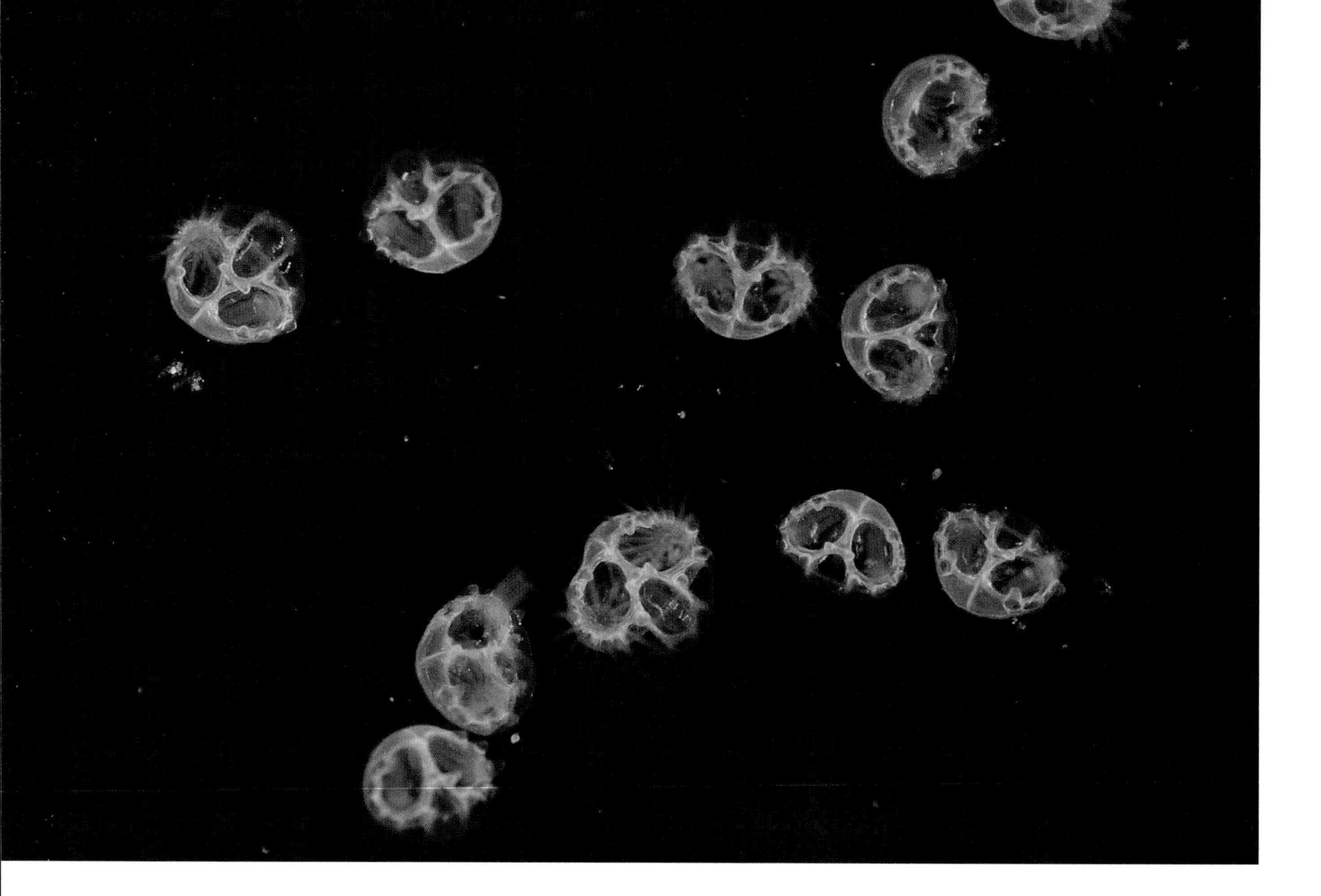

Twin ancestrulas of *Membranipora*, about 1 mm, probably 1–2 days old. On giant kelp blade from Monterey Bay.

The twin-shelled larvae of *Membranipora,* known as cyphonautes, feed in the plankton for several weeks, usually in surface waters, before they settle and metamorphose. The larvae appear regularly in plankton tows; while drifting freely they fall prey to many pelagic invertebrate predators as well as to fishes such as the señorita and the kelp surfperch.

Here the cyphonautes have settled on a blade of giant kelp and metamorphosed into the characteristic twinned zooid ancestrulas of *Membranipora* (the ancestrulas of most other bryozoans are single). Some of these ancestrulas have already extended their feeding tentacles.

Young *Membranipora* colony, 4 mm. On giant kelp blade from Monterey Bay.

From its twinned ancestrula, this young colony is growing by budding. Within some three weeks, such colonies may spread from this tiny beginning to sheets that completely encase sections of kelp blades and floats; they can add a row of zooids every eighteen hours. Here, the crowns of feeding tentacles (lophophores) lie retracted within each zooid. Note the spines at the corners on some of the zooid "houses." Infrequently, colonies with spines may occur naturally; more often the spines are absent. However, in most colonies lacking spines, the spines will develop as a "predator-induced defense"—a response to predation by a nudibranch such as *Doridella steinbergae* (see below). Before this induced defense was recognized, spined colonies of *Membranipora membranacea* were thought to be a wholly different species. The phenomenon has been studied in a laboratory at Friday Harbor, Washington. The response will follow a one-hour exposure to water that bears chemical clues of these predators; it ceases if the predators are removed (Harvell 1984, 1990, 1998). The spines greatly impede predation by these predators. Inducible structural defenses occur most often in clonal or colonial animals, as we saw earlier, for example, in two species of anemones (see *Anthopleura elegantissima* and *Metridium senile*).

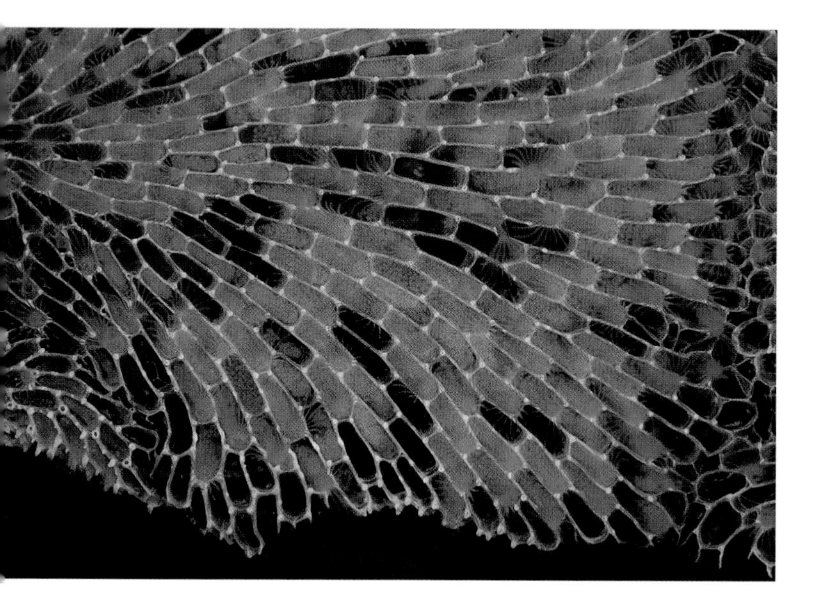

Membranipora colony with ripe eggs, 15 mm. On giant kelp blade from Monterey Bay.

Here the zooids are orange with ripe eggs. The color is derived from the carotenoid pigments of ingested diatoms.

Opposite: Membranipora feeding and spawning, about 5 mm. On giant kelp blade from Monterey Bay.

In this view the lophophores are extended to feed, and at the same time some zooids are spawning. The lophophoral tentacles surround the mouth, which leads to a U-shaped gut. The anus opens just outside the lophophore. A lot is going on here, but most of it is revealed only under a dissecting microscope. At intervals, small groups of adjacent zooids suddenly retract their lophophores in unison and then slowly reextend them, suggesting a neural connection between zooids. Spawned zygotes (fertilized eggs), seen here as tiny white discs, emerge from the inter-tentacular organ (the ITO, a short tube lying between two tentacles at the lophophore's rim). They bounce from one lophophore to another as they are caught and then rejected in the feeding currents of different zooids until they finally reach open water at the edge of the colony.

The sexual reproduction of *Membranipora* has been worked out in fascinating detail (Temkin 1994). Zooids are hermaphroditic. Sperm are released by testes into a zooid's body cavity, or coelom, and emitted in packets, tail first, through the pore in one of two specialized tentacles. As they emerge from the tentacles, many of the sperm packets are swept into exhalent currents and carried away from the colony for possible cross-fertilization in other colonies. Other sperm packets may be drawn into lophophores of sister zooids in the same colony. In either case, the sperm wiggle vigorously amid the lophophore's tentacles; a few may be eaten, but many somehow enter the ITO head first and pass into the recipient's coelom. It isn't known if a chemo-attractant is involved. Almost 100 percent of eggs are fertilized as they enter the coelom through the oviduct. The resultant zygotes then enter the inner end of the ITO, pass out into the lophophore's discharge current, and soon are in the open sea. Embryonic development does not commence until four days after the zygotes are released.

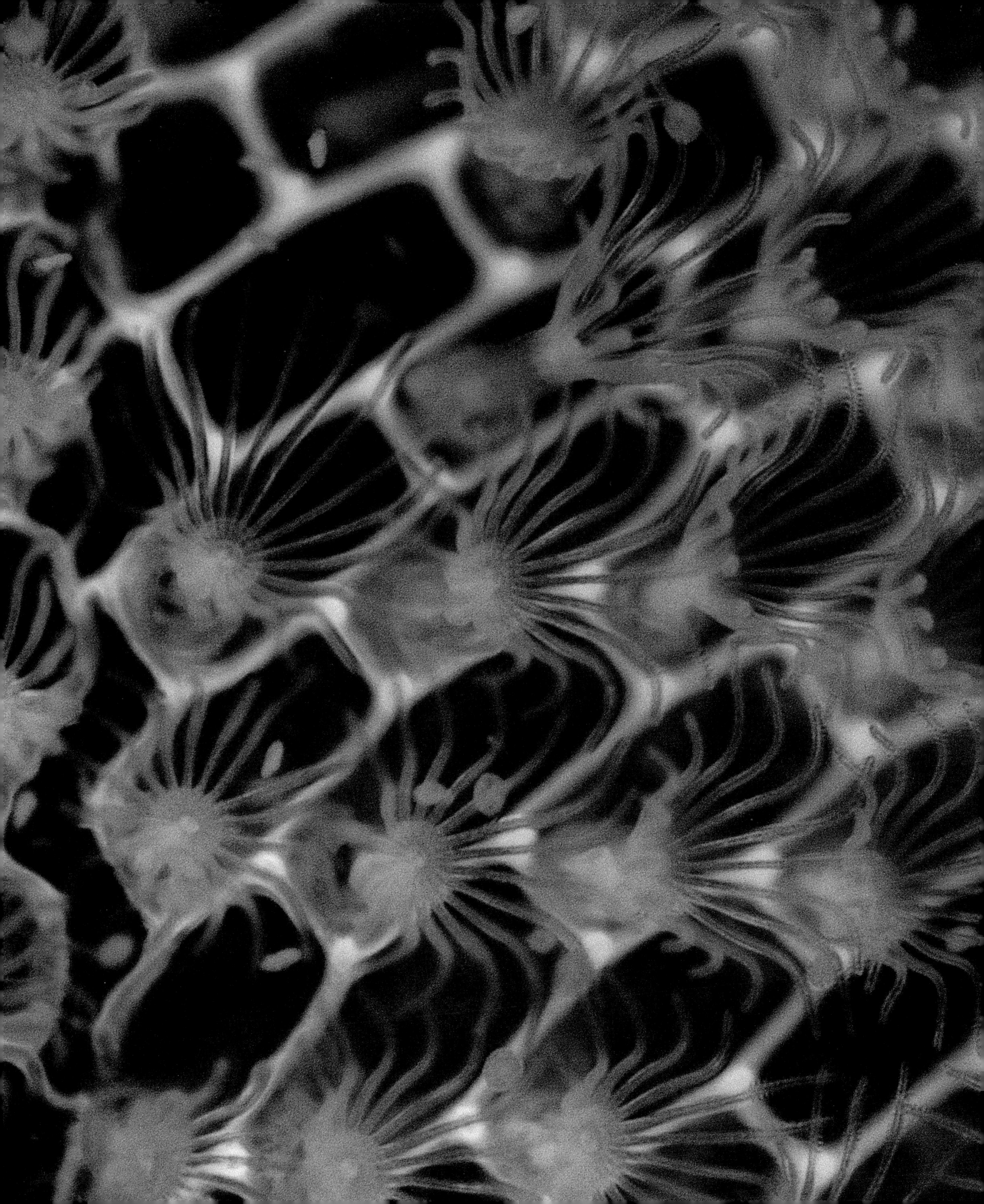

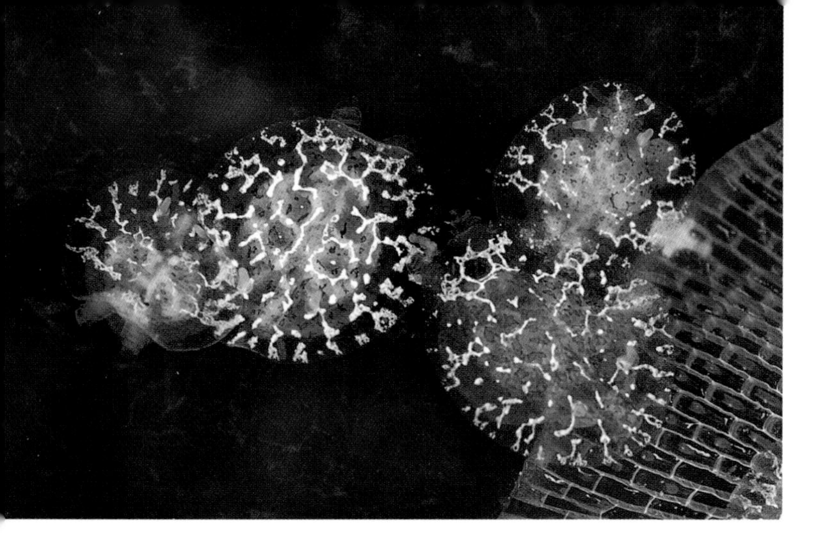

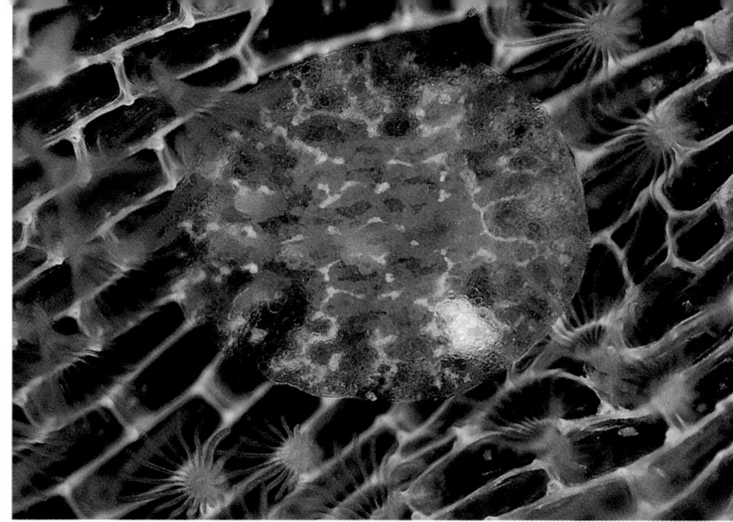

Doridella steinbergae
Phylum Mollusca / Class Gastropoda

Dorid nudibranchs next to the bryozoan *Membranipora membranacea,* field of view 12 mm. On giant kelp blade from Monterey Bay.

These tiny nudibranchs occur only in association with their bryozoan prey, *Membranipora membranacea,* and so are almost always found on blades of giant kelp. The largest nudibranch here measures only about 4 mm long. The pair on the left, oriented right shoulder to right shoulder, are mating, so sexual maturity evidently is reached at a very small size. Because of its minuteness and close resemblance to its prey, we find *Doridella* only by examining kelp blades closely in our laboratory.

Above right: Doridella, about 4 mm long, on its bryozoan prey. On giant kelp blade from Monterey Bay.

When feeding, *Doridella* makes a tight seal with its mouth over an individual bryozoan zooid; uses its rasping tongue, or radula, to perforate the tough membrane overlying the zooid; and sucks out the soft parts. In the laboratory a 6 mm animal consumed about 150 zooids per day (Seed 1976)—an enormous depredation. Note that *Membranipora*'s zooids continue feeding even as *Doridella* rasps out the adjacent ones. The phenomenon surprises us. Based on our observations of the zooids, we believe that they must have neural connections.

Doridella is a "dorid" nudibranch, not an eolid like *Aeolidia.* Most dorid nudibranchs have a circle of gills surrounding the anus, which opens on the back; in *Doridella,* however, the gills lie between the back and the foot, and they may protrude posteriorly, as here. Many dorids feed on sponges, thereby picking up noxious chemicals that may provide a defense. But *Doridella* is almost perfectly camouflaged when feeding on its prey, and this may serve as an adequate defense, at least against visual predators like fish.

Membranipora colony and adjacent *Doridella* egg string, field of view 5 mm. On giant kelp blade from Monterey Bay.

After mating by reciprocal copulation, the hermaphroditic snails deposit these comma-shaped egg masses once or twice daily, each with up to two thousand fertilized eggs. An embryonic period of about eight days follows, and then the larvae hatch. They spend several weeks swimming and feeding on phytoplankton before settling and metamorphosing in the presence of *Membranipora*. Part of the planktonic larva's gut is specialized to sort food particles such as diatoms; at metamorphosis, the gut alters to deal with the semiliquid tissue of *Membranipora*'s zooids (Bickell and Chia 1979; Bickell, Chia, and Crawford 1981). Other changes that occur at metamorphosis include the juvenile nudibranch's withdrawal from its larval shell and its development of a radula. After settlement and metamorphosis, *Doridella* lives for only about three weeks—what a brief, eventful life!

Several of the zooids in this colony appear healthy, for they are feeding and laden with eggs, but the "houses" at the colony's periphery are broken and empty. Additionally, these houses bear outwardly directed, long, translucent spines. The adjacent comma-shaped *Doridella* egg mass is the giveaway. *Doridella* has been there feeding, as well as egg-laying, and the bryozoan colony has responded with its defensive spines. In the laboratory, *Doridella* nudibranchs eat zooids in spined colonies at only about one-fifth the rate at which they eat them in unspined colonies (Harvell 1984).

(Another small dorid nudibranch, *Corambe pacifica*, closely resembles *Doridella* in both appearance and feeding behavior; it too looks virtually like the bryozoan.)

Polycera atra

Phylum Mollusca / Class Gastropoda

Below: A mating pair of black dorid nudibranchs, about 6 mm long. On giant kelp blade from Monterey Bay.

Polycera is still another predator of bryozoans. We have found it associated with *Membranipora membranacea*, as seen here, in the canopy of giant kelp off Del Monte Beach, Monterey. The voracious nudibranchs are said to consume 30–40 percent of their body weight in food daily. Like *Doridella*, *Polycera* rips open the *Membranipora* zooids with its radula and sucks out the soft parts, sometimes even consuming the basal portion of the encrusting colony. The nudibranch's cylindrical fecal pellets are composed of fragmented colony skeletons cemented together with mucus (MacGinitie and MacGinitie 1968). These nudibranchs appear on kelp blades

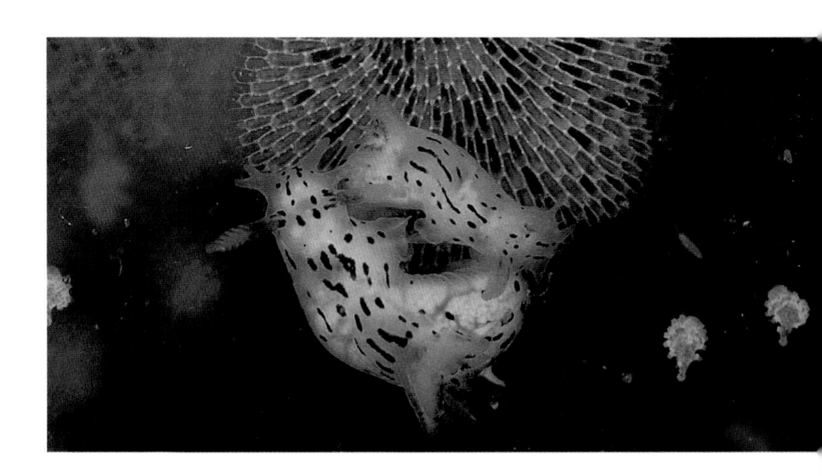

in the spring when new *Membranipora* colonies are forming.

This intimate scene shows the two hermaphrodites as they conclude mutual copulation. Over a period of several minutes we observed as they tightly united, partially disengaged, reunited, and then disengaged. The penis and the vagina lie adjacent to each other in a genital aperture on each nudibranch's right side. During copulation, the needlelike penile process everts, and sperm emerge from its tip. After copulation, each partner stores its newly received sperm in a female receptacle, from which it will fertilize its own eggs. Note that both these nudibranchs are ripe with ova. Following their fertilization, each nudibranch will extrude its eggs through an orifice near the vagina and deposit a characteristic dorid egg ribbon.

Tubulipora pacifica

Phylum Bryozoa / Class Stenolaemata

Opposite: Tubed bryozoan colonies, field of view 12 mm. On giant kelp blade from Monterey Bay.

These bryozoan colonies are small and fan-shaped; their zooids live in white, calcified "houses," each curved upward and 2–3 mm in height. Blades of kelp with this bryozoan growing on them feel like very coarse sandpaper. The zooids' feeding lophophores extend from each active tube; when the zooids retract, they seal their tubes with a terminal membrane. Despite being heavily calcified, the colonies do fall prey to fish.

Although as a whole a *Tubulipora* colony is hermaphroditic, the eggs and sperm develop in separate zooids. Here and there in a *Tubulipora* colony a female zooid, of which there are but a few, will develop a small ovary, and an enlarged brood chamber forms. There are five such chambers in the largest colony here. Sperm are released from the tentacle tips of male zooids. After fertilization via the pore at the base of the female lophophore, a single egg within the brood chamber begins development, and the female zooid disintegrates. The embryo repeatedly divides (a process known as polyembryony), eventually filling the chamber with a hundred or more genetically identical embryos that develop into a clone of ciliated larvae (Hayward and Ryland 1985). Although polyembryony occurs in other bryozoans in the same class and in a large number of insects, it is unusual in the animal kingdom.

In this photo the brood chambers, nestled between tubes, can be recognized by their relatively large apertures. Look closely: one of them is releasing tiny spherical larvae. The nonfeeding larvae have a very short free-swimming life, just seconds to minutes. Upon settling, a larva metamorphoses into a flat disc from which the colony's founder zooid (the ancestrula) projects up as a tiny curved tube—one is at the lower left edge of the photo. In Monterey Bay, we have found *Tubulipora* only on blades of giant kelp off Del Monte Beach.

Tubulipora colonies and a juvenile seastar, field of view 15 mm. On giant kelp blade from Monterey Bay.

This photo was taken in July, evidently a very good time for this bryozoan to reproduce and disperse. Note the ancestrulas and the young colonies of various ages in this extraordinarily dense array. The larger colonies have not yet developed their own brood chambers.

The little seastar shown here measured only 1 mm across. Asteroid larvae feed in the plankton for months before settling, presumably choosing a suitable substratum before they do so. This is the only juvenile seastar that we have found on giant kelp, but in 1986 larvae of the seastar *Pisaster giganteus* settled abundantly on kelp blades in southern California (Herrlinger, Schroeter, and Dixon 1987), and many of the newly metamorphosed youngsters may then have climbed or drifted to the seafloor. The survival odds for the little fellow in the photo would seem rather poor, given the predatory poking-about of the señorita fish, *Oxyjulis californica.*

Pleustes depressa

Phylum Arthropoda / Class Crustacea

The amphipod mimic and its snail model, *Alia carinata* (= *Mitrella carinata*), field of view 16 mm. On giant kelp blade from canopy off Del Monte Beach, Monterey Bay.

Look closely and you will see not only tiny (3 mm) young *Alia carinata* snails among the bryozoan colonies *Tubulipora pacifica*, but also the amphipod *Pleustes depressa*, which uncannily resembles its snail companions (Carter and Behrens 1980). What relationship lies behind the mimicry? Amphipods are sought after by visual predators such as fish. These same fish, however, find many snails unpalatable and tend to avoid them on sight. Additionally, if ingested, many snails can close their operculum and pass undigested through the fish's gut, making them less desirable as prey (Norton 1988). This species of amphipod may better survive amid its predators by closely resembling its less desirable snail model.

Such a resemblance is called Batesian mimicry, after Henry Bates, who first described it for butterflies in 1862. He believed that a palatable species, such as the viceroy butterfly (the mimic) could better survive by closely resembling a very toxic species, in this case the monarch (the model). Here we can see this same evolutionary play going on in the sea. Perhaps pattern mimicry, striking as it may be, is developmentally more efficient or physiologically less costly for the mimic than simply producing its own toxin—a difficult comparison to investigate rigorously!

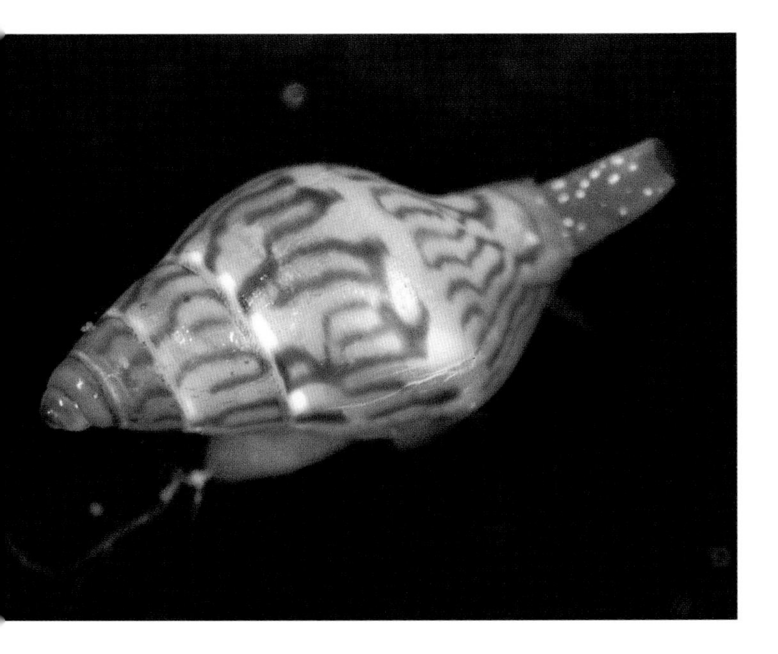

Alia carinata, the snail model, about 3 mm high. On giant kelp blade from Monterey Bay.

In some seasons we find young *Alia* snails abundantly on kelp blades, especially just above the holdfast; adults may live on algae, on rocks, and on surfgrass. The snail appears to be a tiny carnivore and a detritus feeder. Older individuals, 1 cm high, develop a keel-like ridged shell that may help protect them from being crushed by crabs. After internal fertilization, females lay egg capsules from which minute juveniles hatch and crawl away.

Pleustes depressa, the amphipod mimic, about 4 mm long. On giant kelp blade from Monterey Bay.

This amphipod even mimics the behavior of the little juvenile snail, wobbling from side to side as it moves. *Pleustes* probably is a scavenger and detritus feeder. This one is clinging to the bryozoan *Celleporella hyalina* (=*Hippothoa hyalina*). Females retain their internally fertilized eggs in a brood chamber from which the young eventually emerge and crawl away.

As we have just noted, adult *Alia carinata* snails look very different from the juveniles; in addition to having a shell with a well-developed keel, they are dark on one end, not light. Another species of this amphipod genus, *P. platypa,* mimics not the juveniles but rather these adult *Alia* snails (Crane 1969)!

Plumularia sp.
Phylum Cnidaria / Class Hydrozoa

Hydroid colony, field of view 38 mm. On giant kelp blade from Monterey Bay.

Hydroid colonies are among the several colonial marine animals that used to be called zoophytes, or "animal plants," for they combine a plantlike appearance with their animal traits. This hydroid looks like a tiny bush, but like other hydroids it is in fact a colonial predator whose feeding polyps capture tiny zooplankton. The scattered colonies of *Plumularia* hurry through their growth on the short-lived kelp blades. They do not release free-swimming medusae but instead retain that sexual stage of the life cycle within the colony as medusoids. The medusoids produce either sperm or eggs, depending on the sex of the whole colony. Embryos develop into ciliated larvae (planulae) that are released and, after a bout of swimming, settle and metamorphose into the attached polyp stage.

Fronds of male *Plumularia* colony with reproductive polyps, feeding polyps, and defensive polyps, field of view 5 mm. On giant kelp blade from Monterey Bay.

The elongate, white structures seen in this picture contain testes; released sperm ripen on the granular, golden central core (Hyman 1940). No trace remains of the medusoids that produced these structures. The feeding polyps, meanwhile, have extended their nematocyst-laden tentacles. And the scattered tiny white nubbins are mouthless defensive polyps, also loaded with nematocysts.

Opposite: Frond of female *Plumularia* colony with reproductive polyps and medusoids, field of view 5 mm. On giant kelp blade from Monterey Bay.

Arising from the main stalk, these reproductive polyps contain medusoids with eggs in various stages of maturation. Although the medusoids are greatly reduced compared to free-swimming medusae, they are fully sexual. The densely packed white eggs are ripe and await fertilization by sperm from a male colony.

Each type of hydroid polyp is structurally simple, but the combination of three types of polyps (feeding, defensive, reproductive) makes the entire colony a complex entity, and it is even more complex when it bears a medusoid stage of the life cycle.

Eucopella (= Campanularia) sp.
Phylum Cnidaria / Class Hydrozoa

Hydroid colony showing expanded feeding polyps, field of view 9 mm. On giant kelp blade from Monterey Bay.

This colonial hydroid, too, grows commonly on kelp blades. A proliferating network of stolons, or tubular connecting runners, gives off feeding polyps and medusoid-bearing polyps. The feeding polyps and their expanded tentacles, laden with nematocysts—well shown here—are borne on upright stalks about 2 mm high. These carnivorous polyps prey on tiny zooplankton. Donald P. Abbott and some of his students once explored the behavior of the feeding polyps (1987): A gentle touch near the polyp's mouth with a probe wet with saliva, and the mouth suddenly opens. A firm touch to the tentacles, and both the upper and lower tentacles retract. A gentle touch with a clean probe on one upper tentacle, and the upper tentacles close over the mouth while the lower ones remain inactive. Touch the tentacles on one side of the polyp with food, and they will bend toward the mouth and even right into the mouth, while those on the opposite side are inactive. These complex behaviors suggest multiple neural connections in the tiny feeding polyps.

Male *Eucopella* colony showing reproductive polyps, field of view about 5 mm. On giant kelp blade from Monterey Bay.

Here, male medusoids 1 mm high have developed from reproductive polyps. The milky fluid filling some of these medusoids and the stolons consists of sperm that eventually will be released. Well-developed radial canals, clearly visible within some of the medusoids, are extensions of the colony's gut that carry nutrients to the nonfeeding medusoids.

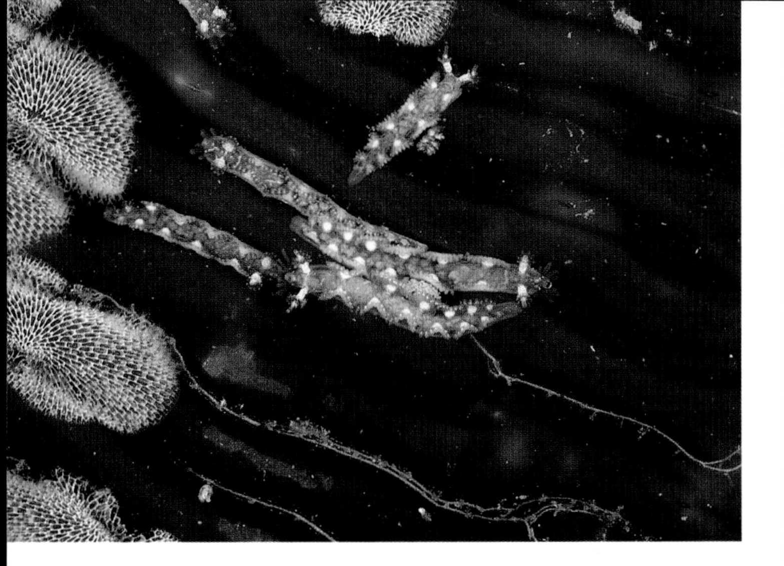

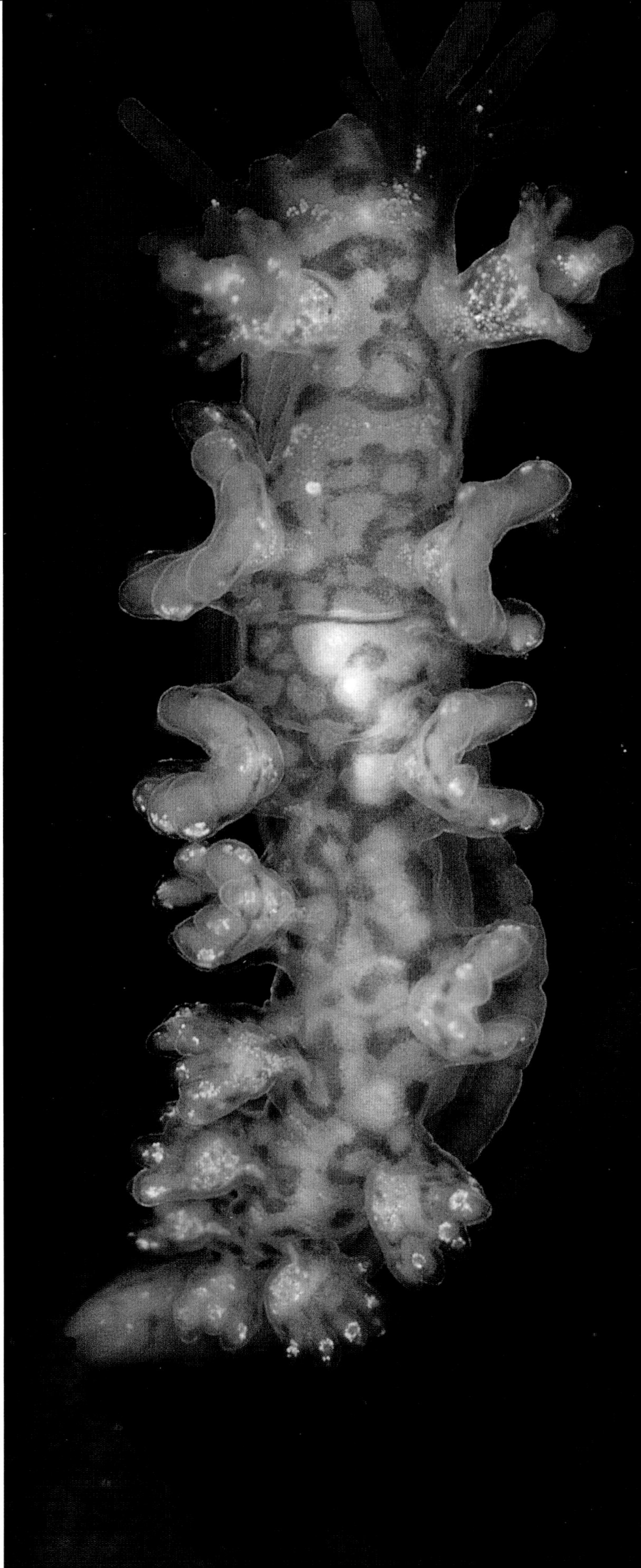

Hancockia californica
Phylum Mollusca / Class Gastropoda

Nudibranchs, field of view 36 mm. On giant kelp blade from Monterey Bay.

Once, while hanging over the edge of our inflatable raft, we noticed what looked like tiny toothpicks lying in the grooves of the floating blades of giant kelp. What a find! The inconspicuous but beautiful little *Hancockia* can easily be overlooked. In this view there are several adults, an egg string between them, and one juvenile. Like many nudibranchs, *Hancockia* is short lived. It often occurs with ephemeral prey such as the hydroid *Eucopella.* Note that several stolons of *Eucopella* colonies have lost their polyps, undoubtedly owing to the nudibranch's depredations. *Hancockia,* in the suborder Dendronotacea, does not store the hydroid's nematocysts in its dorsal cerata as eolid nudibranchs do in theirs. The bryozoan *Membranipora membranacea* here has expanded its lophophores and appears unmolested; *Hancockia* seems not to feed on it.

Right: Close-up of *Hancockia,* about 10 mm long. On giant kelp blade from Monterey Bay.

The cerata, outpouchings of the gut, have a delicate palmate shape in this nudibranch.

Eubranchus rustyus
Phylum Mollusca / Class Gastropoda

Eolid nudibranch, 6 mm long. On giant kelp blade from Monterey Bay.

Here *Eubranchus* is preying on the hydroid *Obelia*. Note the eye spots that lie just behind the nudibranch's sensory rhinophores; they detect light but do not form an image. For guiding behavior, they certainly are not as important to *Eubranchus* as the rhinophores' olfactory function. Like other eolid nudibranchs, *Eubranchus* will store the hydroid's nematocysts in its cerata as a borrowed defense.

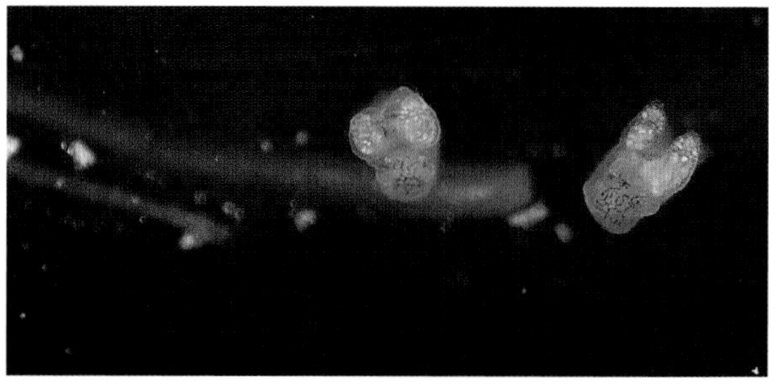

Juvenile nudibranchs, probably *?Eubranchus,* about 0.3 mm long. On giant kelp blade from Monterey Bay.

We can find and observe nudibranchs in their minute juvenile stage only under a dissecting microscope. The individuals here have recently completed metamorphosis from the larval stage and have lost their larval shell and acquired a radula; each possesses one pair of cerata and one pair of rhinophore buds. Here, the juveniles clearly are feeding on a hydroid stolon, probably that of *Obelia.* They may be *Eubranchus* nudibranchs; their cerata look quite similar to those of the adults nearby.

In one species of *Eubranchus* (*E. doriae*), larval metamorphosis is induced by a chemical from its natural hydroid prey (Bahamondes-Rojas and Dherbomez 1990). We can speculate that *Obelia,* too, contains a substance that induces *Eubranchus rustyus* larvae to settle and metamorphose into juveniles. Sophisticated chemical analyses are identifying more and more chemically mediated behaviors in nature that until recently were unexplained or attributed to chance. As E. O. Wilson put it, "Ninety-nine percent of the animals find their way by chemical trails laid over the surface, puffs of odor released into the air or water, and scents diffused out of little hidden glands and into the air downwind" (1992, 4).

Melibe leonina

Phylum Mollusca / Class Gastropoda

Right: Lion nudibranchs on giant kelp, field of view 33 cm. Monterey Bay, 7 m deep.

These large, dramatic animals cluster in great numbers in the kelp canopy off Del Monte Beach in Monterey Bay, where they present a spectacle as they swim from frond to frond, feed, mate, and lay eggs. But sometimes they vanish, an entire generation of the short-lived organisms apparently having died off.

Or did they die? In November 1970, in kelp beds near our own study site off Del Monte Beach, some twenty thousand apparently thriving individuals, after residing there for several months, vanished over a period of only eight days and were never again located (Ajeska and Nybakken 1976). Similarly, seasonal swimming and dispersal have been observed in sexually mature populations in Puget Sound (Mills 1994). To our knowledge, these mysterious comings and goings have never been satisfactorily explained. The apparent migrations of sexually mature individuals would seem to add a method of dispersal in addition to that afforded by *Melibe*'s planktonic larvae.

The kelp crab *Pugettia producta* is one of *Melibe*'s few predators; the nudibranchs rapidly swim away when *Pugettia* approaches. In contrast, fish and some seastars actively avoid the nudibranch (Bickell-Page 1991). When disturbed, *Melibe* releases a defensive secretion (it smells like watermelon) from skin glands concentrated in its cerata and in its oral hood. The glands are encased in muscle and surrounded by ciliated sensory cells. Stimulation of the sensory cilia causes muscle contractions, which in turn force out the chemicals, evidently noxious to many predators.

Melibe swims by rhythmically flexing the body from side to side; at the same time, it tightly closes its oral hood, thereby reducing drag. While swimming, neural impulses from two ganglia fire alternately in a rhythm, driving motor neurons that innervate the flexor swimming muscles.

Melibe leonina, about 7 cm, characteristically perched on a kelp blade. Monterey Bay.

Only *Melibe* nudibranchs have evolved the large oral hood seen here. The adult animals feed by opening and closing the hood like a net that sweeps planktonic crustaceans from the water. The hood is expanded by hydrostatic pressure and closed muscularly. When it closes, the numerous short tentacles on its rim interdigitate and retain the prey—a design rather like the prongs on a Venus flytrap. The mouth then pushes forward to swallow the food. *Melibe* lacks the scraping radula found in most nudibranchs, which is not needed for this mode of feeding on tiny suspended prey. The hood's little "horns," the rhinophores, appear to be chemosensory: when they are removed surgically, *Melibe* no longer shows its usual feeding response to water that has passed over crustaceans. Tactile cues also are important in feeding, since *Melibe* responds to inert particles in the water, but vibrations are important as well: although dead crustaceans elicit a moderate feeding response, live and active ones elicit the maximum response (Watson and Chester 1993).

Below: Egg ribbon of *Melibe,* about 3 cm, on a kelp blade. Monterey Bay.

Among the assemblage of *Melibe* in the kelp canopy are many mating pairs of these hermaphrodites, and egg ribbons are common. The egg capsules form chains within the ribbon's gelatinous matrix; each capsule contains 15–25 eggs, and the entire ribbon may harbor 30,000 eggs. Feeding larvae hatch from this mass and swim away. We have found juvenile *Melibe* in the kelp canopy at Pt. Pinos, at the edge of Monterey Bay, apparently miles from the nearest

adults, which suggests that currents disperse the larvae long distances. A young *Melibe* has a distinctive feeding pattern that is quite different from the adult's. It applies its hood directly to the kelp surface, traps food items such as copepods and tiny amphipods right there, and sucks them into its mouth.

Epitokous polychaete annelid

Phylum Annelida / Class Polychaeta

Right: Syllid polychaete, 1 cm long. On giant kelp blade from Monterey Bay.

This mature polychaete worm is undergoing a yearly transformation to become a sexually reproductive, swimming form called an epitoke—a metamorphosis that takes place in several families of wandering polychaete worms. Annelids, or "ringed worms," are made up of a chain of segmented or repeated body parts and thus are well adapted to develop the detachable, wormlike part known as an epitoke. Often the entire worm is transformed, but in some species, as in the unidentified polychaete (family Syllidae) shown here, only the posterior half becomes an epitoke. The worm's body constricts, and a new head with eyes forms where the parts will soon separate. Here the developing epitoke's appendages have become broad and paddle-like, preparing it for its swimming, pelagic existence. Other changes under way inside the epitoke include a reduction of its musculature and gut, with the energy and space saved given over to the increased production of gametes. In many polychaete species, the anterior portion of the worm, if it survives this extraordinary amputation, regenerates its usual, more mundane posterior portion and eventually may produce more epitokes.

Epitokes develop from both males and females. They are strongly drawn to light. On a summer night, often at or just after a full moon, epitokes of a species will detach synchronously and swim to the surface. Wriggling females burst as they release their eggs while swarming males encircle them with a cloud of sperm. The likelihood of fertilization at the surface is greater than it would be in a midwater encounter. After they discharge their gametes, the epitokes of both sexes die, leaving behind an abundant but vulnerable new generation of polychaete larvae.

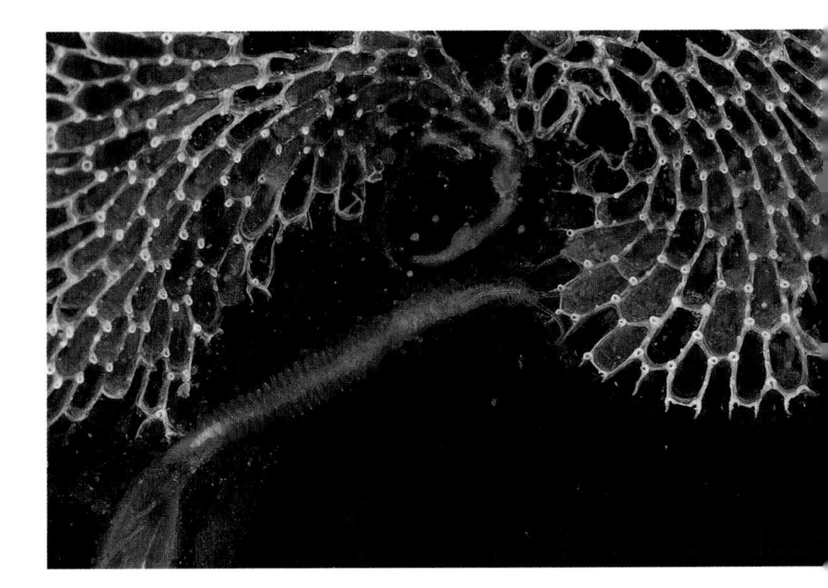

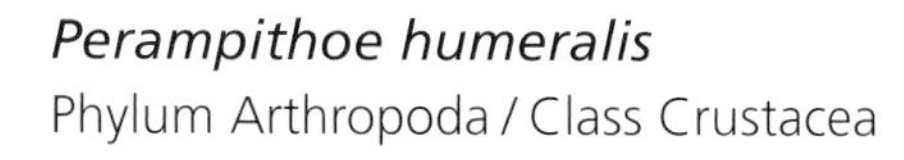

Perampithoe humeralis
Phylum Arthropoda / Class Crustacea

Nest of kelp curler amphipod, field of view about 4 cm. On giant kelp blade from Monterey Bay.

This amphipod folds and glues down the edge of a blade of giant kelp to form a carefully constructed, tunnel-like home and refuge. The inhabitant's antennae can be seen poking out of the nest, and graze marks made by these herbivores are evident on the kelp blade. Often females are found nesting while males evidently cruise from nest to nest in search of receptive mates.

Female *Perampithoe* brooding her young, about 20 mm. On giant kelp blade from Monterey Bay.

We find these amphipods' nests by hanging over the edge of our inflatable boat and closely examining the surface canopy of the kelp forest. One nest contained this female brooding her young. The little ones are hanging head down beneath her thoracic segments. During brooding, the mother protects the young but does not feed them; they apparently survive on their yolk until the mother's next molt forces their release. Development is direct, with no larval stage. On one occasion we detached the juveniles from their mother and found that she was brooding forty-five young at once.

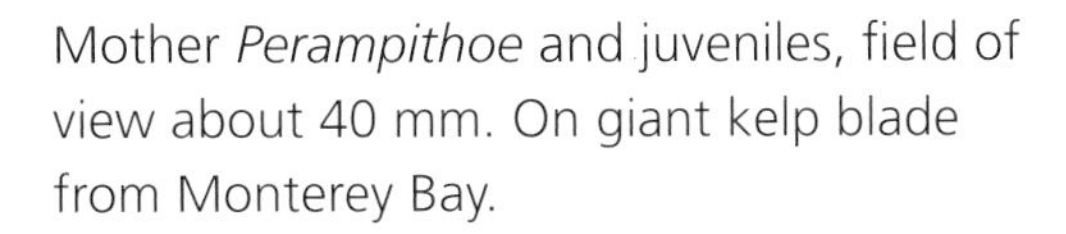

Mother *Perampithoe* and juveniles, field of view about 40 mm. On giant kelp blade from Monterey Bay.

Within the nest, the tiny creatures move about rapidly by sculling on their side. A disturbed occupant will do a very agile flip-flop and scull itself to the opposite end of its tunnel. The broodmates may continue to return to their nest and mother after making feeding excursions outside. Gradually the little herbivores will take on a golden brown color as the kelp's carotenoid pigments suffuse through their bodies. This color camouflage, together with their nesting behavior, may help protect them from predatory fishes.

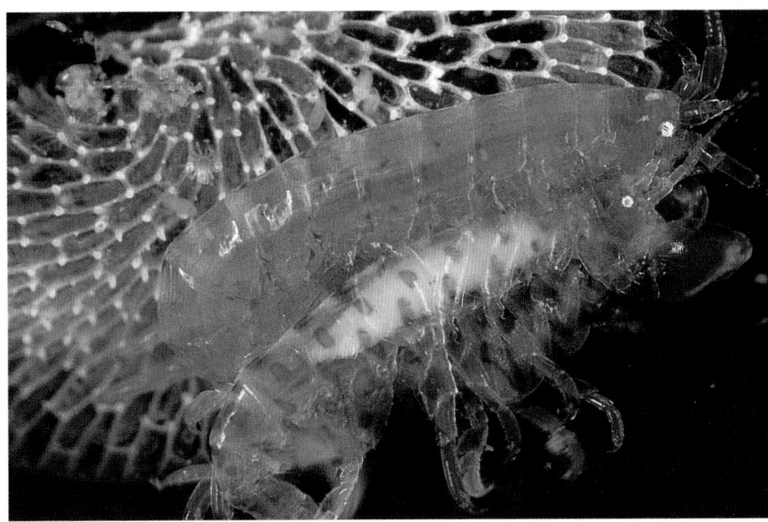

A mating pair of *Perampithoe*, about 20 mm long. On giant kelp blade from Monterey Bay.

This male is clasping a female with his anterior thoracic limbs. He will carry her around beneath him for several days until she molts, at which time he will emit sperm that she sweeps into her marsupium, or brood chamber. She will then release her eggs into her marsupium, where fertilization will occur. Note the white mass of eggs still within the female's body.

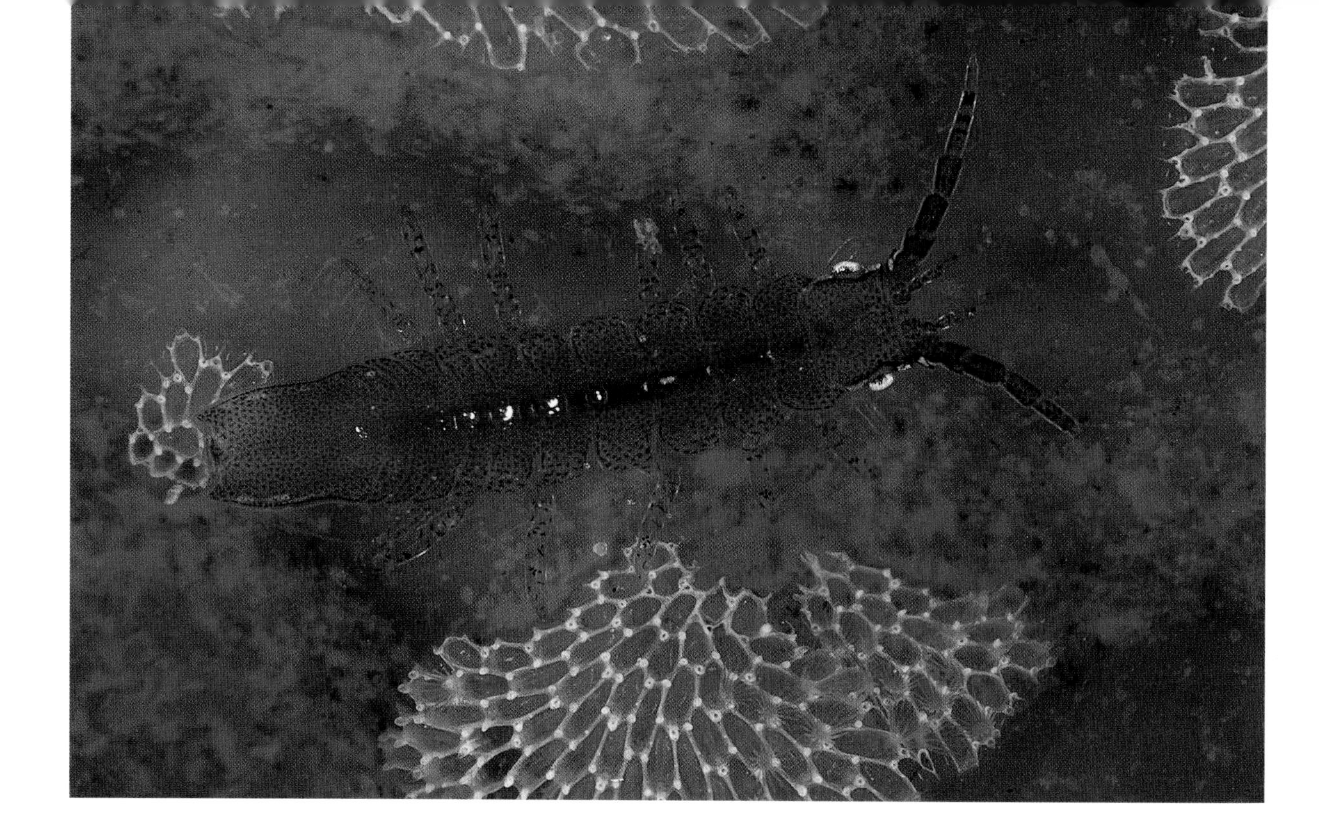

Idotea resecata
Phylum Arthropoda / Class Crustacea

Isopod, about 8 mm long, and bryozoan colonies. On giant kelp blade from Monterey Bay.

These herbivorous isopods, common inhabitants of the kelp forest, are the little crustaceans that we find hooked onto our wet suits when we emerge from the sea after exploring a kelp bed. They graze heavily on giant kelp, particularly near the kelp's floats, where growth of the blades occurs and where the seaweed's nutrients are most concentrated. As a consequence of their grazing, many blades become detached, leaving long sections of stipes denuded. Several species of fish, including the señorita, *Oxyjulis californica*, prey on this isopod. On one occasion, when for some reason control by predators evidently failed, these isopods multiplied in huge numbers (up to 1,060 individuals per blade) and wiped out a large kelp canopy in southern California (Bernstein and Jung 1979).

The golden brown color of giant kelp is due primarily to the accessory photosynthetic carotenoid pigment fucoxanthin. Intuitively one might expect *Idotea* to achieve its nearly perfect color camouflage by depositing this pigment in its cuticle. Not so. Instead, the isopod synthesizes new compounds from the carotenoids of kelp (as does its relative *Idotea montereyensis*) and deposits a green pigment in the inner layer of cuticle and red in the outer one. Together, these two pigments blend to match the brown of *Idotea*'s favorite food (W. Lee and Gilchrist 1972). A population of green *Idotea resecata* (green in both cuticle layers) occurs on eelgrass, *Zostera marina*, but apparently it is isolated from the population on kelp and has lost the capability for color change.

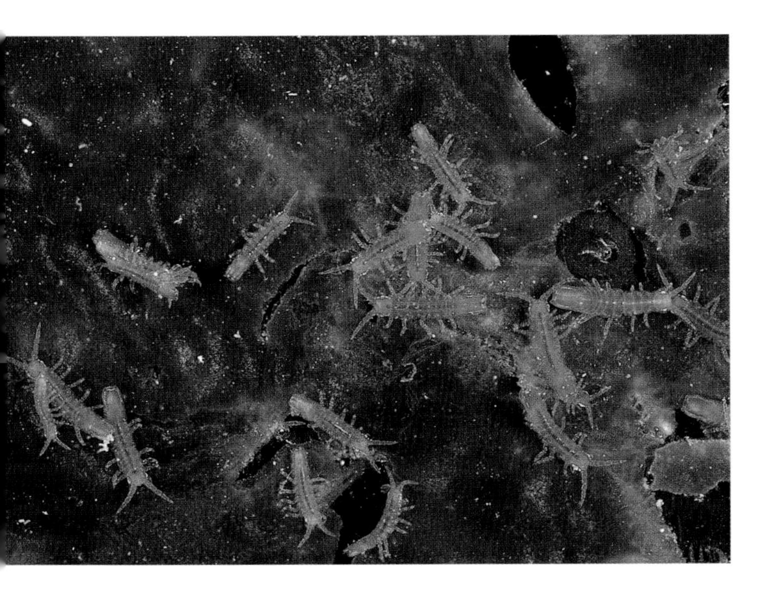

Juvenile *Ideotea* isopods, field of view about 2.5 cm. On giant kelp blade from Monterey Bay.

As far as we know, *Idotea* does not make nests. Fertilization is internal, and females brood their young until they bear what at first look like little adults minus one pair of legs (which the young gain in the course of molts toward adulthood). These young have not yet dispersed, and they appear to lack at least that pair of legs.

Notice that there are holes in the kelp blades and streaks where its outer layer has been grazed off, leaving a lighter background against which the lighter young isopods blend in. This damage is from the isopods' grazing, comparable to the terrestrial clues of insect herbivory one finds in the garden.

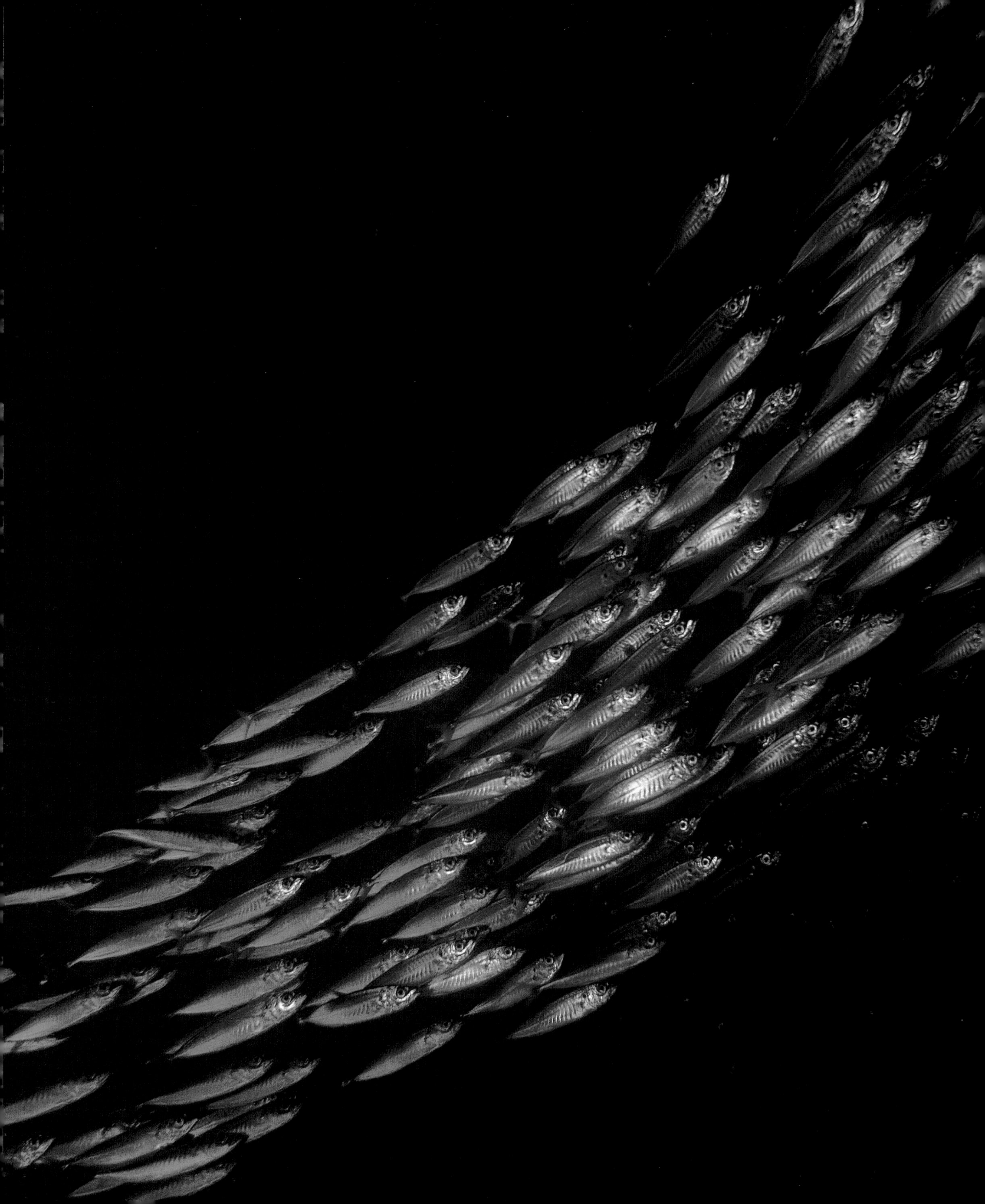

4

The Outer Bay

The outer bay's sea surface shimmers above, but as an open body of water it has no apparent sides or bottom. Nevertheless, subtle and often well-defined boundaries do shape the communities of organisms that live there. They are determined by currents, the compensation depth, nutrient levels, reduced oxygen levels, temperature profiles, salinity patterns. Many of these features dramatically shift as the year progresses through the spring and summer's coastal upwelling to the fall's oceanic period. Organisms ranging in size from microscopic to the world's largest animal, the blue whale, are forever swimming, or drifting, or migrating vertically up and down in this three-dimensional world. They are difficult to study in their natural habitat, and many of the methods used today date from the 1800s—trawls, tows, or dredges that miss agile organisms and damage delicate ones. Although open water organisms' undisturbed interactions and adaptations still elude us to a large degree, today satellite imaging, collecting and photographing by ROVs (remotely operated vehicles), and recording by echosounders, together with more traditional methods, are expanding and sometimes altering our understanding of the outer bay.

When contrasted to the enormously productive shallower inshore waters, offshore oceanic waters often appear relatively barren. Away from upwelling, offshore waters are somewhat warmer, nutrient poor, relatively clear, and without the dense blooms of phytoplankton and zooplankton that can be sampled in plankton nets inshore. But there is a world of tinier organisms (<60 microns) in the outer bay and beyond that pass through these nets (Pomeroy 1974; Fenchel 1988), and their rate of respiration may exceed that of the larger "net" plankton by a factor of 10. Indeed, as much as 90 percent of photosynthetic activity in oceanic waters may be due to these organisms, particularly cyanobacteria, which are visible only through a powerful microscope. So, as Pomeroy puts it, the net plankton, far from being the grasses of the sea, are in fact the sequoias of the sea, while most photosynthesis is done by the phytoplankton that slip through nets. In addition to these pho-

School of jack mackerel *(Trachurus symmetricus).*

tosynthetic organisms, there is a world of tiny consumers (e.g., nonphotosynthetic flagellates and ciliates) that are likewise too small to be recovered in towed nets.

And it doesn't stop there. Only in the 1990s have we learned that even tinier organisms are a significant force in the ocean's ecology—specifically, viruses, twenty-three *billion* of which would be required to make a visible particle (Suttle 1999). Viruses, we now know, are far more numerous in the world's oceans than even cyanobacteria, for example. Viruses live and replicate only by infecting living cells. To attain their huge numbers, on any given day they must infect, and destroy, equally huge numbers of both phytoplankton and zooplankton. How the ocean's food web paradigm is changing!

What about organisms that *are* recovered in nets? At the outer edge of the bay, trawls of plankton nets from the surface to depths of a thousand meters recover vast numbers of small organisms, predominantly ones that spend their entire lives in the sea—copepods, shrimp, krill (euphausiids), siphonophores, and small fish. At night, many of these organisms are vertical migrators that live by day in the oxygen minimum zone (OMZ) at depths of 500 to 1000 m. At night they rise to the sea's surface to graze on sunlight-dependent phytoplankton and on zooplankton in relative safety from larger visual predators such as fish.

The oxygen content in the OMZ is greatly reduced and may even approach zero. This low content is attributed to the respiration of the organisms living there, to the absence of oxygen production from photosynthesis at these dark depths, and to the bacterial decomposition of sinking organic matter. At even greater depths, the oxygen content rises somewhat owing to the influx of oxygen-rich cold water that sank from the surface in far-away high latitudes and flowed south to our coast.

During World War II, warships used sonar echoing equipment to track both submarines and the sea bottom. The submarine search often was frustrated by perplexing midwater reflections termed "deep scattering layers," under which an enemy sub might lurk undiscovered. At the time, these echoes were attributed to some physical discontinuity in the water, but now we know that shoals of midwater animals reflect sonar and create a perceived false bottom to the sea. In synchrony with the animal communities' migrations, scattering layers move toward the surface at night and start their descent at dawn or with a rising full moon. Today fishermen use their ship's sonic depth-finder to locate schooling fish, for the fishes' swim bladders are particularly efficient at reflecting sonic waves.

More conspicuously, oceanic waters are populated by shoals of relatively large gelatinous animals—medusae, comb jellies, salps, pelagic snails—as well as powerfully swimming nekton—migrating cetaceans, schools of pelagic fishes, squids. The occasional appearance of such oceanic animals close within our bay can be dramatic. While diving in kelp beds, we have been thrilled to see a great gray whale lumber by, almost within arm's reach. In Carmel Bay, we often witness a feeding frenzy when marauding sea lions drive schools of anchovy to the surface and pelicans plunge into the sea, mouths agape, to scoop up the prey. In the fall, onshore winds and currents sweep beautiful medusae and other gelatinous organisms into the bay. These animals are ephemeral, and within days or weeks they disappear. But they remind us of the vast oceanic world around us.

Balaenoptera musculus
Phylum Chordata / Subphylum Vertebrata / Class Mammalia

Blue whale, 21 m long. Central California beach.

Opposite: Baleen plate with bristles, about 50 cm.

California has an estimated population of some 2,200 blue whales, the largest animals alive today on the planet, and every year in the late summer and early fall some of them congregate off central California and Monterey Bay at the edge of the continental shelf. They sometimes have a yellow coating of diatoms, hence their historical common name of "sulphur bottom." Why are they here, and why at this time of year?

The answer lies in their diet. Blue whales feed only on tiny shrimplike creatures called krill or euphausiids, small crustaceans they strain from the sea with their bristled baleen plates. To understand krill is to understand the blue whale. Every July in this area, a generation of krill reaches maturity, and their behavior then makes them vulnerable to the whales' exploitation. At night the krill migrate vertically and disperse in shallow water to feed on phytoplankton, thereby escaping visual predators and presenting a very poor, diffuse target for the whales. During the day, however, they aggregate in dense masses at depths of 150–200 m on the edge of the Monterey Canyon, as we have learned with echosounders. And whales, it turns out, dive to precisely these dense krill masses, where they take the krill with huge gulps, perhaps a thousand krill at a gulp, thirteen gulps a dive, 120 dives a day—all told, a thousand kilograms of krill per day (Croll 1998). What a contrast with the lifestyle of a toothed whale such as *Orca*, a predator of marine mammals.

Our photo shows one of these leviathans—a young male, possibly a year old—lying on its back on a central California beach, its right pectoral fin extending over author Libby. Blocks of baleen have fallen out on the sand. A blue whale's baleen consists of some two hundred plates of fingernail-like keratin, nestled together and set transversely into the upper jaw. Each plate has a filtering fringe of long bristles, as shown here.

Blue whales feed with huge gulps. Their mandible drops almost vertically, and their throat pleats, clearly seen in the photo, expand as the mouth distends with water and krill. Then the whale closes its mouth while thrusting its tongue forward; the mouthful of water exits through the baleen, and the krill are retained. How extraordinary that possibly the largest animal that has ever existed on the planet is at the top of such a very short food chain: phytoplankton—krill—blue whale!

In September, when krill become abundant off Baja California, the blue whales move south in pursuit. It has been suggested that their immense size may be due to the need to store enough energy to move rapidly for long distances without feeding. The California population travels as far as Central America in their search for the patchy, widely scattered krill. With large size there is also relatively less skin friction to impair motion and relatively less heat loss.

The problem with the notion that immense size is needed to store energy for long-distance migrations is that ruby-throated hummingbirds also migrate long distances—annually across 600 miles of water from Florida and Georgia to Mexico. They, too, store fat before their migration; they are aided, however, by winds and by "refueling" while in transit (Terres 1980).

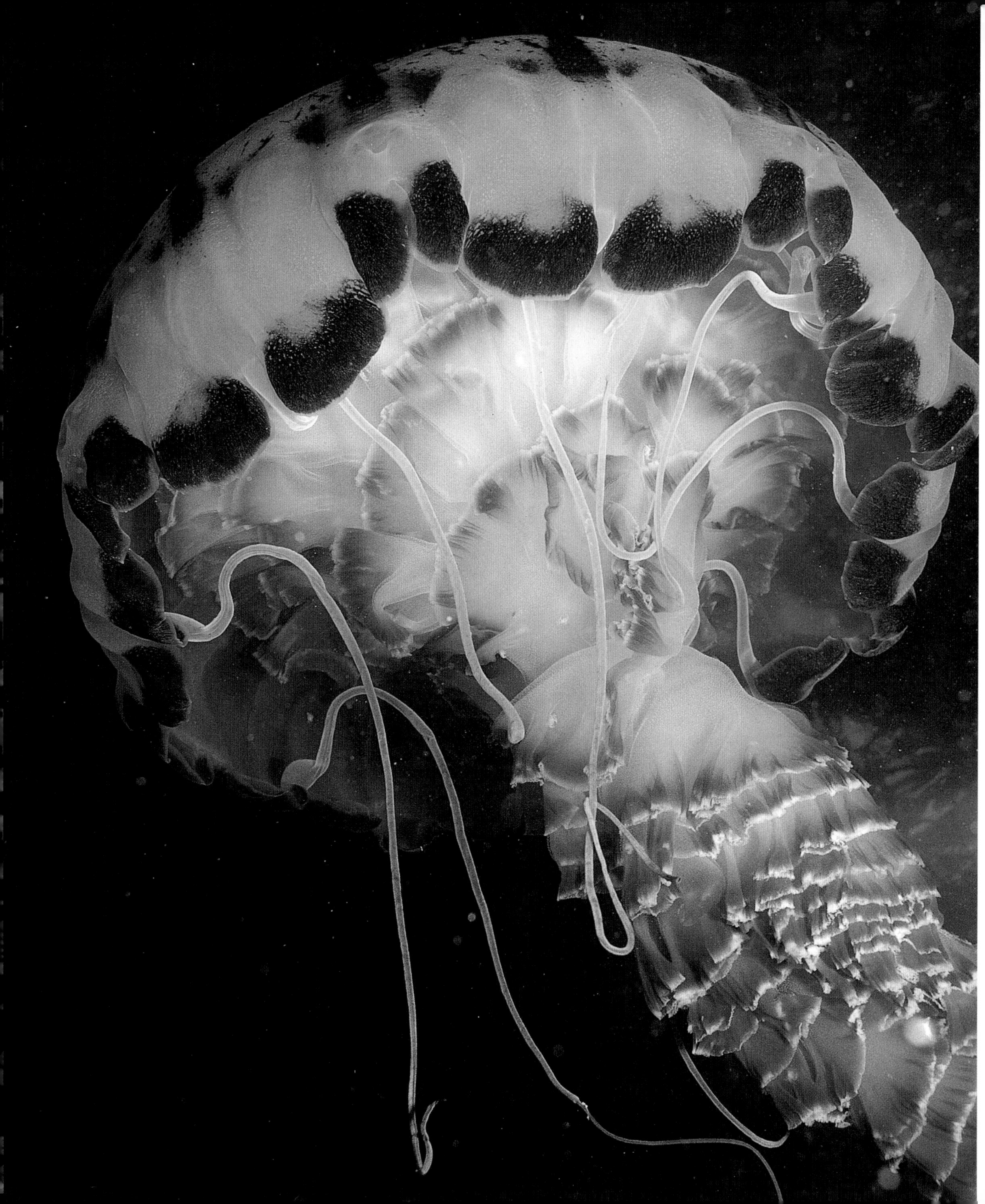

Pelagia colorata
Phylum Cnidaria / Class Scyphozoa

Opposite: Purple-striped jellyfish, 26 cm in diameter. Monterey Bay, 5 m deep.

"Jellyfish," yes, but fish no. It is more telling to call these cnidarians medusae. This name echoes the Medusa of Greek mythology, one of three Gorgon sisters, who, deprived of her charm, found her hair changed to hissing serpents. Her aspect became so frightening that those who looked at her turned to stone. Do dangling tentacles suggest writhing serpents? In a way they do, as the tentacles in *Pelagia* are loaded with venomous nematocysts. Medusae use these stinging cells to subdue their prey. They advance with a pulsating rhythm and trail long, frilly, spirally coiled, deadly mouth lobes as well as tentacles. Medusae have separate sexes; their gonads lie in the gut wall beneath their bell.

Entangled *Pelagia,* 25 cm in diameter. Monterey Bay.

This medusa has reached the end of its voyage, shipwrecked in the kelp canopy. Most of its tentacles and mouth lobes are missing, perhaps eaten by blue rockfish. Within Monterey Bay we have seen other *Pelagia* being torn apart by blues. It isn't clear how fish avoid (or tolerate) the medusa's nematocysts—not only the tentacles but also the mouth lobes and even the exterior of the bell are covered with these potent weapons. Harbor seals and the ocean sunfish, *Mola mola,* also feed on these medusae. Don't touch *Pelagia* with your bare hands even when it lies stranded, for its nematocysts may still sting painfully.

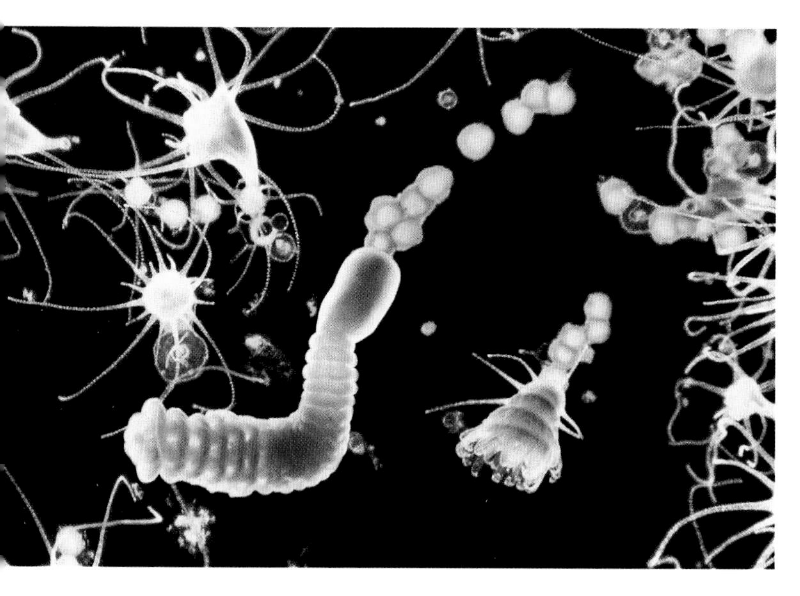

Pelagia polyps, 10 mm high, releasing tiny medusae. (Photo courtesy of Freya Sommer)

Many cnidarian organisms have an alternate form in their life cycle, such as the polyps and medusae of the hydroid *Obelia* (see Fig. 10). Until recently, however, *Pelagia* was not known to have the attached polyp stage common to many medusae (Fig. 11). But in 1988 the biologist Freya Sommer, while on an offshore field trip, put several freshly caught medusae in a large tub. By the time she returned to shore, the stressed animals had released gametes, and fertilization led to the production of planula larvae. After a week of culture, the larvae settled and metamorphosed into tiny polyps. Later, the polyps elongated and budded new polyps in turn by producing "pedal cysts," in this photo the yellow blobs at the polyp's point of attachment. After about three months of culture, the polyps segmented to resemble a stack of plates (a process called strobilation, well seen here in the central polyp) and their tentacles resorbed into the trunk. Still later, the uppermost plates successively detached as tiny medusae, called ephyrae, and swam away (Sommer 1988). Thereafter, *Pelagia*'s long-lived polyps periodically regenerate their tentacles and again feed, and then strobilate again.

The polyp stage of *Pelagia colorata* has not been found in the wild, even though we now know what to look for. The medusae, mysteriously, have a very limited range: only from southern Oregon to northern Baja California. *Pelagia*'s polyps may be restricted to some very localized and certainly obscure site that awaits discovery by an enterprising—or fortunate—biologist.

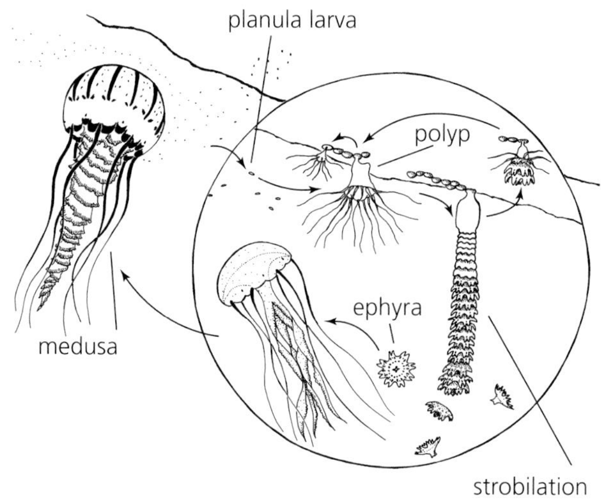

Fig. 11. *Pelagia colorata* life cycle

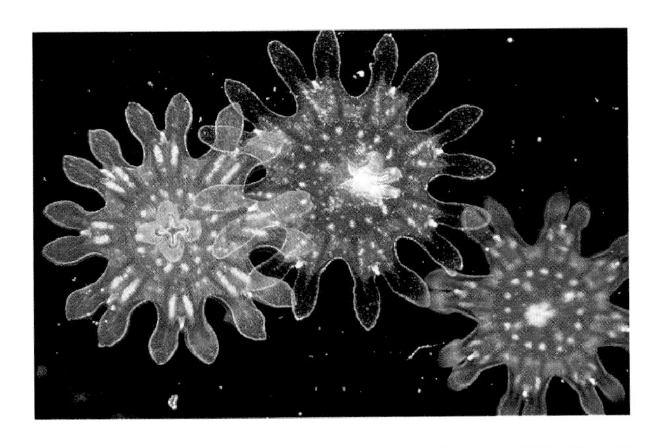

Pelagia ephyrae, 3 mm across. (Photo courtesy of Freya Sommer)

What a thrill it must have been to watch these tiny tentacleless medusae being released! Documenting the reproductive cycle of *Pelagia colorata* was a triumph of careful observation and technical know-how.

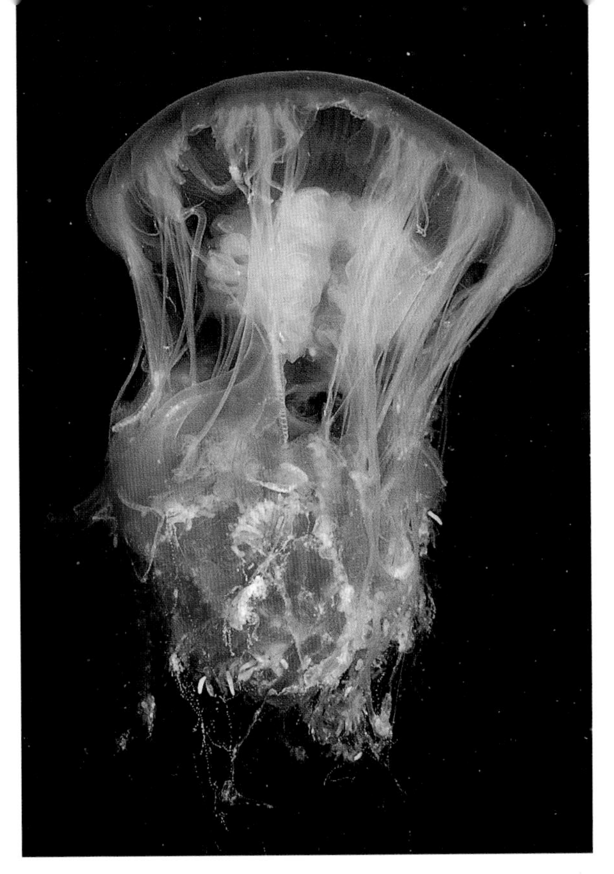

Phacellophora camtschatica

Phylum Cnidaria / Class Scyphozoa

Above right: Egg-yolk jellyfish, about 35 cm high. Carmel Bay, 7 m deep.

Phacellophora may occur offshore in huge numbers, but they are found in Monterey Bay only when unusual winds and surface ocean currents drive them in, usually in the fall. This medusa's prominent egg-yolk orange feeding lobes hang down below the light yellow gonads and gut. The medusa has caught another gelatinous animal, the siphonophore *Praya* sp., reduced now to some stringy material with tiny yellow bodies entangled by the tentacles beneath the medusa's mouth lobes. *Phacellophora* preys not only on other gelatinous organisms such as small medusae, ctenophores, and siphonophores, but also on crustaceans.

Phacellophora medusae the size of this one swim by contracting the bell at about ten beats per minute. They often fish by making slow vertical excursions during which they rise with their muscular power stroke (as in this photo) and sink while motionless or beating slowly. At other times they may swim horizontally and suddenly reverse directions, swimming back through their array of already deployed tentacles. During these maneuvers the marginal tentacles may splay out to form a large net. The tentacles are capable of enormous elongation, up to seventeen times the bell's diameter. Rapidly swimming prey like the moon jelly, *Aurelia aurita*, get snared in *Phacellophora*'s deployed net, while smaller, weakly swimming prey are caught in vortices in the wake of the medusa's power stroke, swept against the long, trailing, venomous tentacles, and captured. In Puget Sound, a huge *Phacellophora* medusa with a bell diameter of 45 cm was observed with several thousand small prey amid its tentacles, in its oral feeding arms, and in its stomach (Strand and Hammer 1988). Ocean sunfish, *Mola mola*, eat these medusae, as do sea turtles.

Phacellophora polyp, about 4 mm in diameter, releasing an ephyra. (Photo courtesy of Freya Sommer)

This laboratory-reared polyp is about to bud off a tiny medusa. And stacked like saucers immediately beneath it are four more well-differentiated medusae, each awaiting its turn to break free. The polyp's feeding tentacles have been resorbed. What a graphic illustration of cloning this extraordinary picture is, with its wealth of detail! The polyps of *Phacellophora,* like those of *Pelagia,* have never been found in the wild.

Phacellophora ephyra, about 6 mm across. (Photo courtesy of Freya Sommer)

After they detach, ephyrae pulsate away by contracting their bell with a rapid beat. Note that the bell of the *Phacellophora* medusa has 16 marginal lobes, twice the number of *Pelagia.* The medusa's sensory structures, located at the bifurcation of each lobe, appear here as discrete white spots. How do these infants feed before their tentacles develop? They capture food particles with one of the marginal lobes and bring them to the mouth by flexing the lobe. And they already are well defended: the scattered bright speckles are little patches of nematocysts.

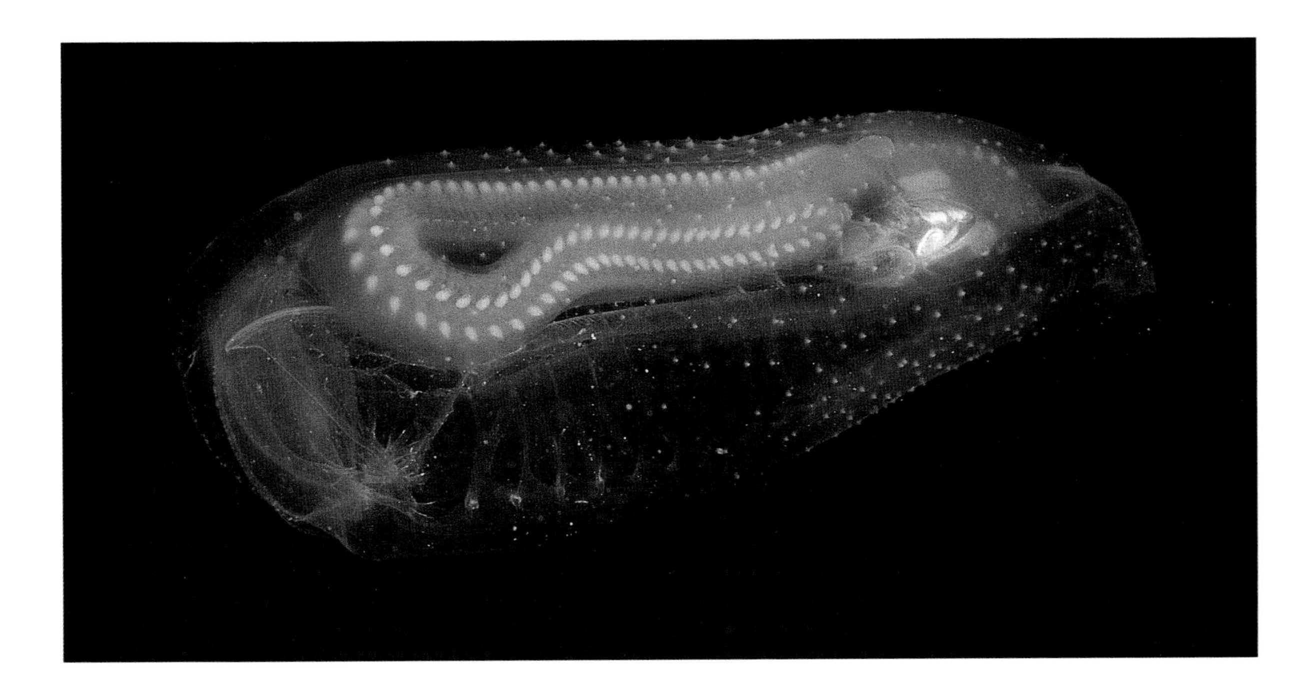

Thetys vagina
Phylum Chordata / Subphylum Tunicata / Class Thaliacea

Salp, solitary phase, 10 cm long. Carmel Bay.

This large body is the solitary phase in the life cycle of a salp, a pelagic tunicate. These gelatinous oceanic animals only occasionally are swept into our bays. The solitary phase gives rise to an "aggregate" phase by growing a ropelike stolon that subdivides into a double chain of gonad-bearing little cloned bodies. Each chain may consist of fifty to several hundred clone-mates at a time. A single big solitary individual may eventually produce hundreds to thousands of these little progeny. Parent and progeny are all genetically identical, yet the two forms are so different that at one time they were given separate names.

Its transparent tunic reveals this young animal's developing stolon and its gill-lined pharynx. Salps jet-propel themselves by contracting circular muscle bands to force water through their barrel-shaped body. The incurrent (oral) and excurrent (cloacal) apertures of these animals accordingly lie, jetlike, at opposite ends of the body. With the water flow of their jet propulsion, salps feed, respire, dispose of wastes, and disperse sperm. Feeding and locomotion go on continuously. They feed by passing water through a constantly produced mucus sheet that coats the inner wall of the pharynx. Periodically the food-clogged mucus sheet is rolled into a cord and carried to the esophagus by ciliary action. Captured particles, ranging in size from a millimeter down to less than a micron, are often smaller than even tiny animals such as copepods can exploit. Salp fecal material includes bacteria, the small fecal pellets of other grazers, skeletons of copepods, diatoms, and other small organisms. In many parts of the sea the salps' fecal pellets are so abundant they play an important role in the cycling of organic and inorganic materials to the deep sea and seafloor (Madin 1974; Alldredge and Madin 1982).

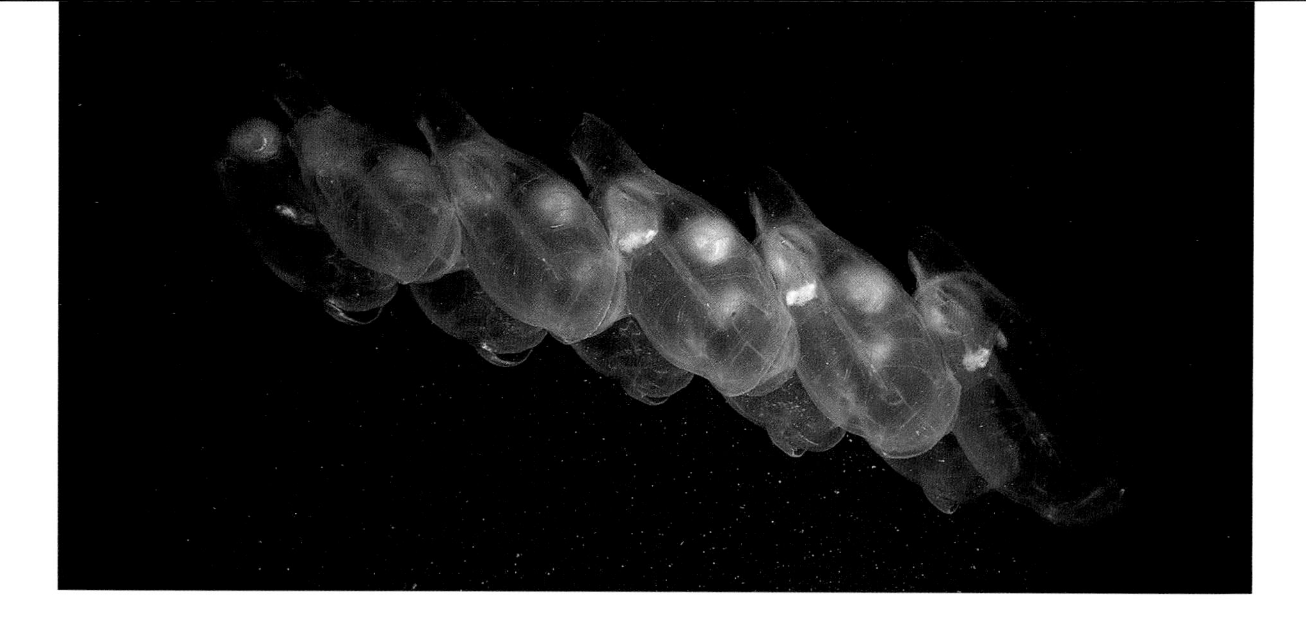

Salpa sp.
Phylum Chordata / Subphylum Tunicata / Class Thaliacea

Pelagic tunicates, "aggregate" phase, each about 2 cm long. Carmel Bay.

Long chains of these little zooids, later to become sexually mature salps, are budded off from the stolons of the pelagic tunicate's big solitary phase. Cloning by a single solitary animal such as *Thetys* may produce chains of hundreds of zooids that are oriented in a double chain, as seen here. Each zooid is slightly asymmetrical, so back to back they fit snuggly together and produce a smooth, streamlined chain. Adjacent zooids in the chains have neural connections that transmit sensory information to each zooid's "brain," where nerve impulses for swimming originate (P. Anderson and Bone 1980). The chains swim forward smoothly as the zooids pump water through their pharynges, though their contractions are not coordinated. A tactile, photic, or chemical stimulation will bring about a change in behavior. When the zooid in front is touched slightly, all zooids suddenly begin pumping simultaneously, and the chain will swim backward; a touch on the last zooid, and the chain will instantly move forward (Bone, Anderson, and Pulsford 1980). Stronger stimulation, or the ocean's turbulence, can soon break up the chains.

The cloning and a short generation time of salps result in almost inconceivably large numbers of these zooids. A single swarm off southern California in 1950 covered 3,500 square miles with an average density of 275 salps/m^3 to 70 m depth (Madin 1974). They have even stopped a ship by choking the filters of the cooling water intake.

As grazers on phytoplankton, these tunicates have a major impact on the oceanic food web. Perhaps most significantly, they compete for food with tiny larval fish, and when very abundant they can greatly reduce the subsequent recruitment of juvenile fish. In addition to reducing their food supply, an abundance of salps increases fishes' foraging time, thus exposing them to more predation. A few gelatinous organisms, some larger fishes, and birds are known to take salps.

Each zooid in the chain is a sequential hermaphrodite. The few eggs of the initially female salp are fertilized by sperm released by an older male. The parent broods its embryos and nourishes them through a placenta-like membrane until they reach a size permitting survival on their own. Each embryo will become a solitary-phase animal directly, without passing through the tadpole larval stage so characteristic of benthic tunicates.

Lepas anatifera
Phylum Arthropoda / Class Crustacea

Pelagic barnacles, to 18 cm long, on fishing float. Monterey Bay Aquarium.

A detached fishing float is a made-to-order habitat for these seagoing barnacles. So different from the rugged *Pollicipes polymerus* of our surf-swept shores, delicate *Lepas* settles on floating objects such as logs, bottles, ships, and, as in this example, fishing floats. Note the dense clustering of the animals; the long, naked, vulnerable stalk; and the extended, feathery thoracic legs. These appendages rhythmically gather in suspended food particles and may enhance respiration by creating water currents around the body. *Lepas* feeds on midsized planktonic organisms; in the laboratory it will even take animals larger than itself. The barnacle's cyprid larvae (the settling stage) sense light and pressure, cues that enable most of them to stay in the top 10 cm of the sea, facilitating encounters with floating objects (Walker 1995).

This barnacle's especially naked stalk can extend as long as 50 cm, and its protective plates are lightened and reduced relative to other barnacles, making it buoyant but also vulnerable to predators. It forms persistent populations only in transient pelagic habitats away from the pressures of benthic life (Foster 1987). Even in its preferred habitats, both the larvae and the adults are taken by fish and seabirds. *Lepas* barnacles seem to be an ancient stock that has found a last refuge in the open sea.

Velella velella
Phylum Cnidaria / Class Hydrozoa

By-the-wind sailor, about 6.5 cm across. Carmel Beach.

It's a rite of spring: when strong winds blow from the south or west, millions of these little oceanic hydrozoans, adrift at that season on the sea surface, are cast ashore in windrows, forming a blue fringe along our beaches. Our photo of two stranded individuals shows the "sail," a vertical, cantilevered, chitinous sheet that is angled to the axis of the "hull." Drifting before the wind, the animals veer about 45 degrees to the right of the prevailing northwesterlies. Along our coast this is normally enough to keep them offshore, but southerly or very strong onshore winds will bring them toward land. Interestingly, the set of the sails of animals off the coast of temperate Asia mirrors that of our *Velella*s, so that the northwesterlies—which prevail there too—blow the animals to the left and so keep them at sea as effectively as the winds do here.

Velella is designed for stability and seaworthiness (Francis 1985, 1991). The sail itself is a marvel of engineering ingenuity: triangular, slightly thicker at its base, stiffened by superficial thickened ridges, and still quite flexible. This design permits smooth bending when under load, allows recoil when the wind lets up, and minimizes the risk of kinking. The whole animal tilts when under sail, hull broadside to the flow of oncoming water.

These enigmatic creatures are variously interpreted as a single elaborate polyp floating upside down with its mouths and tentacles dangling into the sea, or as an upside down, polymorphic colony with a single, central, feeding polyp surrounded by reproductive polyps with mouths. Concentric air chambers in the body wall act as a float. *Velella* feeds on relatively passive prey such as fish eggs and crustacean larvae. The polyp harbors symbiotic single-celled algae (zooxanthellae), which, through photosynthesis, may contribute to its nutrition (Banaszak, Iglesias-Prieto, and Trench 1993).

Velella's vibrant color arises from the ingested prey's carotenoid pigments, which are modified to become blue and serve to screen excess light in the very bright sea-surface world.

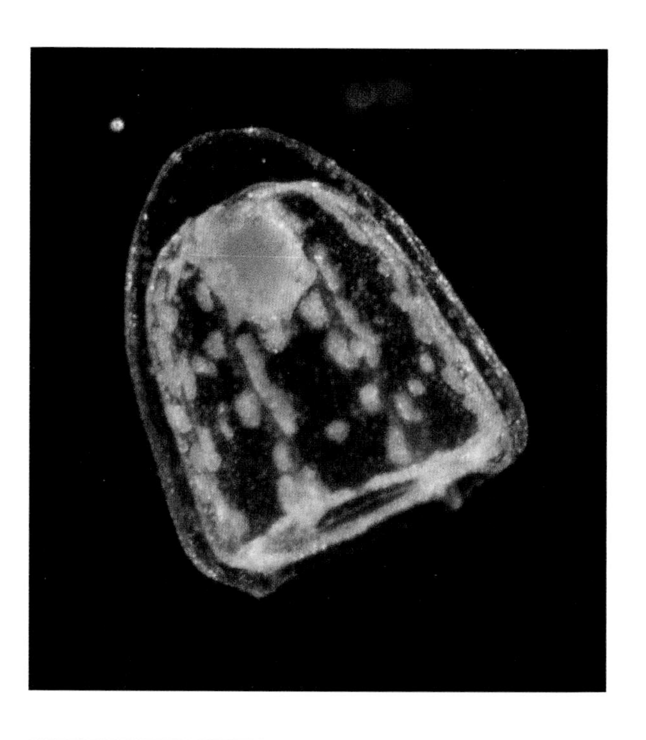

Velella's female medusa stage, 2 mm across. (Photo courtesy of Freya Sommer)

When reproductive, the by-the-wind-sailor buds off tiny medusae from small stalked mouths beneath its float that surround the polyp's larger central mouth (Fig. 12). In nature, *Velella*'s medusa stage has been found with regularity only in the Mediterranean (Larson 1980), but now the medusae can easily be obtained by culture. The medusae, like the adult polyps, bear zooxanthellae; in this photo they appear as yellow granules. Although the medusae can be induced to feed, under lights having a full solar spectrum they will develop to maturity without feeding, deriving nutrition, presumably, from their zooxanthellae. Female medusae each produce one relatively large (0.5 mm) orange egg, as seen here inside the upper part of the bell. When the eggs are fertilized by a male medusa's sperm, tiny orange planula larvae can be obtained and kept alive for many months. In the laboratory, the larvae could not be induced to metamorphose into the next stage (Freya Sommer,

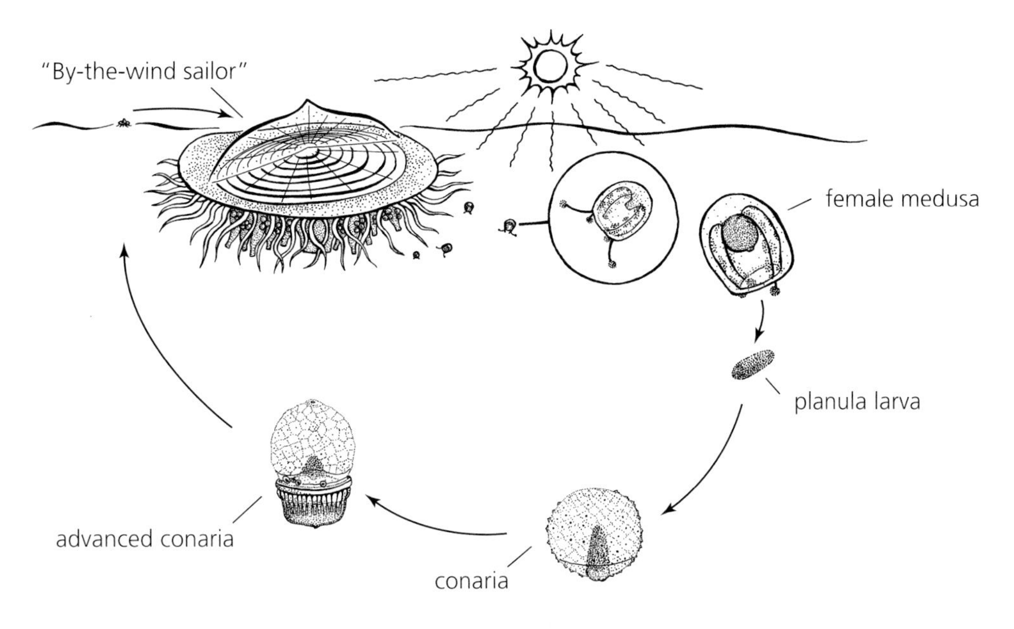

Fig. 12. *Velella velella* life cycle

pers. comm.). It is instead thanks to R. Woltereck, who around the turn of the century was working in a Russian zoological laboratory at Villefranche on the southern French coast, that we know the next part of the story (Woltereck 1904). On winter trawls in the Mediterranean Sea he recovered *Velella*'s next larval stage, called the conaria, from depths of 600–1000 m. So it seems that *Velella*'s medusae descend to these depths and produce planulae that then develop into conaria. It is puzzling that the medusa and conaria stages of an animal that produces vast flotillas on the world's warmer seas have apparently been found only in the Mediterranean and once in the North Sea (Bieri 1977; Larson 1980).

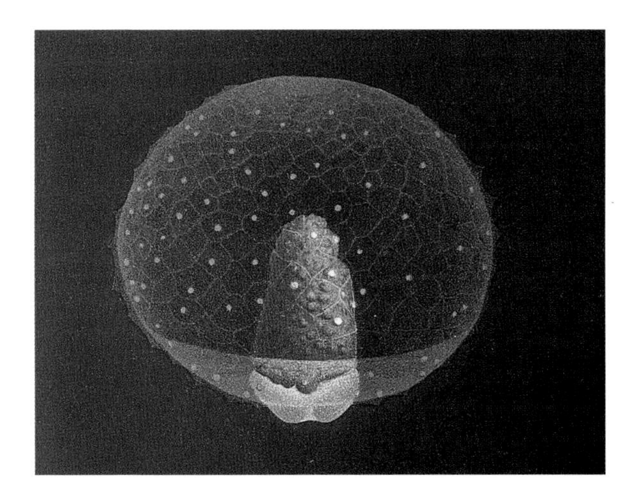

Velella's conaria larval stage, about 1 mm across. (Photograph of Woltereck's 1904 drawing from life)

Woltereck observed that as this tiny tentacleless conaria develops, the red cone hanging in its vesicular cavity diminishes in size and an aperture opens at the base of the sphere: the larva's future mouth. Almost certainly, the red cone is a remnant planula larva that developed into the conaria. During its one to four months of wintertime residence at depth, the conaria enters still another larval stage, developing a rudimentary sail and floats, secreting a drop of oil, becoming buoyant, and by spring ascending to the sea's surface. Bieri (1977) observed these 1–2 mm floating larvae off Baja California. He described them as nearly spherical *Velella*s with no obvious asymmetry in the sail orientation. To him they resembled minute air bubbles.

In general, *Velella* appears to spend several months at the sea surface during spring plankton blooms, at which time the animals sail in the wind feeding while their symbionts photosynthesize in the sun. Then *Velella*'s medusae descend to depth and go through the conarial and later larval development, producing another brood that will ascend to the surface in synchrony with the fall plankton bloom. Individuals of the fall brood of sailors are smaller than those in the spring, as might be expected since plankton blooms then are so much smaller. It's a carefully tuned strategy that apparently prevents undue dispersal of the animals while at depth and ensures that they will be able to take advantage of favorable surface conditions.

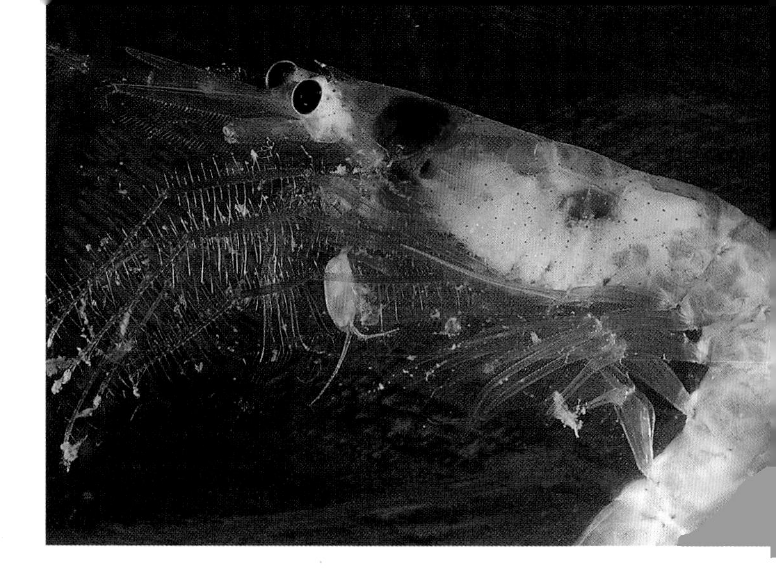

Sergestes, probably *similis*

Phylum Arthropoda / Class Crustacea

Pacific sergestid shrimp, 25 mm long. From 850 m depth, daytime trawl, Monterey Bay.

This little *Sergestes* shrimp was brought to the surface in a midwater plankton-net trawl, accompanied by copepods, amphipods, krill (euphausiids), hydromedusae, comb jellies (ctenophores), and lantern fish (family Myctophidae). Together, these organisms make up a "deep scattering layer," recognized by ships' depth sounders. By day *Sergestes* lives in these deep, cold, oxygen-poor waters, but at night it migrates hundreds of meters vertically into surface waters. What purpose is served by such a demanding regimen? Near the surface at night *Sergestes* finds prey such as herbivorous copepods, numerous there because they feed by day on the sunlight-dependent phytoplankton. *Sergestes*'s large stalked eyes may help its nocturnal foraging. And at night, of course, the shrimp are less vulnerable to many of their visual predators, such as albacore and yellowtail rockfish. Another benefit of vertical migration: as they move up and down, the shrimp may encounter ocean currents that carry them this way and that horizontally, and eventually to new and greener pastures above. At night the shrimp can occur in such concentrations at the surface—up to 4,500/m^3—that, in the

North Pacific whale feeding areas, sei and fin baleen whales find them a plentiful source of nutrition (Butler 1980). Massed like this, these shrimp could even support a commercial fishery.

Careful studies of *Sergestes* from deep-diving subs have painted a detailed picture of the shrimp's behavior and remarkable adaptations (Cowles 1994). This shrimp's second antennae (extending beyond the photo frame) are fully three body lengths long. The antennae are sensitive to tactile stimuli, and they may be chemosensory—useful attributes in the search for food (such as the copepod entrapped in this shrimp's thoracic appendages). However, the antennae remain extended during swimming, clearly a source of drag. The shrimps swim by continuously beating five pairs of abdominal appendages, called swimmerets, clearly seen in this photo. They move along at about 7.4 cm/sec except to pause for grooming. To reduce drag, the shrimp folds up its thoracic appendages. To escape from danger, *Sergestes* rapidly flexes its abdomen and tail fan, abruptly propelling itself backward. The powerful abdominal muscles used for this maneuver are the tasty morsels we enjoy in our salads. *Sergestes* also camouflages itself by ventrally emitting a dim bioluminescence that perfectly balances the intensity and spectrum of the light that penetrates the midwater depths inhabited by this little shrimp during the day. The visual pigment in its eyes is tuned to detect the appropriate spectrum, thus mediating the defensive reaction (S. M. Lindsay et al. 1999). This sophisticated adaptation may render *Sergestes*' silhouette almost undetectable to predators that hunt by sight.

Gaussia princeps
Phylum Arthropoda / Class Crustacea

Female black prince copepod, ventral view, about 10 mm long. (Specimen courtesy of Karen Light, Monterey Bay Aquarium)

This copepod has turned on her back, a characteristic copepod behavior in the laboratory. Her extremely long first antennae are characteristic of copepods. Copepod food products are stored in a midgut oil sac, seen in the photo as a red streak. The spherical dark brown bodies are stored sperm packets.

Most copepods, tiny to almost microscopic planktonic crustaceans, are extraordinarily numerous and a critical link in the marine food web of surface waters. They feed on phytoplankton and in turn are fed upon by larger animals such as other crustaceans and fish. But *Gaussia* is different. For one thing, during the day it lives in the oxygen minimum zone (OMZ) at depths down to 1100 m. For another, you don't need a microscope to see this "giant": at ten times or more the size of most of its copepod relatives, it exhibits the remarkable phenomenon of "abyssal gigantism," whereby the normally tiny crustaceans of shallow waters are replaced by relatively huge species in deep waters. Perhaps gigantism is not what one might expect in a region characterized by darkness, cold water, reduced oxygen concentrations, the absence of phytoplankton, and sparse food of any kind. Under these conditions, however, growth is slow, sexual maturity delayed, and predation likely reduced. Such factors may permit a longer life and the attainment of larger size (Nybakken 1997). In an apparent contradiction, however, the OMZ also harbors fishes that are tiny, such as the grotesque and relatively huge-mouthed angler fish, which measures only 8 cm—an example of "abyssal dwarfism."

It takes unusual adaptations to live in the oxygen minimum zone. There, *Gaussia* can reduce its metabolic rate to a very low level and live anaerobically to compensate for the greatly reduced ambient oxygen concentration, only 1/25 that present in shallow water. In fact, *Gaussia* can survive for as long as fourteen hours without any oxygen and possibly for several weeks after only one meal (Childress 1975). In contrast to many copepods it is omnivorous, and it even may catch other zooplankton, scarce though they may be at depth. During *Gaussia*'s daily sojourn in the OMZ, however, the oxygen debt in its tissues builds up and the little crustacean may get hungry. At night, therefore, *Gaussia* migrates vertically up to about 200 m depth, where it obtains more oxygen, rids itself of the unwanted products of anaerobic metabolism, and finds more food.

Gaussia is one of many deep-dwelling organisms that produce brilliant bioluminescent displays. Following mechanical, electrical, or ultraviolet light stimulation, *Gaussia* ejects blue glowing material from scattered pores all over its head, appendages, and core body (A. Barnes and Case 1972). The glow remains for a second or two and then diffuses and decays. Meanwhile, *Gaussia* rapidly darts away. Neighboring copepods do not respond to the discharge, but predators may be diverted to the glow and thus away from *Gaussia* itself. Bioluminescence thus appears to provide the copepod with an escape mechanism. William Beebe, whose adventures in his deep-sea bathysphere were among Libby's favorite reading during her youth, observed a shrimp's luminescent display in 1934: "When a shrimp was startled, as when it bumped against the bathysphere window . . . a rocket-like burst of fluid was emitted with such violence that the psychological effect was that of a sudden explosion. . . . For an instant the shrimp would be outlined in its own light—vivid scarlet body, black eyes, long rostrum—and then would vanish, leaving behind the confusing glow of fluid" (Beebe 1934, 304).

This female has successively mated twice, each time with a different male, as males produce only one sperm packet at a time and she has two sperm storage receptacles (Cuoc et al. 1997). The males probably are attracted to receptive females by a pheromone. During mating, a male grasps his mate with a specialized hooked antenna and attaches a sperm packet to her genital area. The packet's contents then discharge into one of her sperm receptacles. The receptacles appear here as spherical dark brown bodies; they contain sperm as well as associated secretions. Since insemination can occur only once in the female's reproductive life, all her eggs eventually will be fertilized by the stored sperm. The eggs are released following fertilization, later to develop into the copepod's first larval stage.

Melanostigma pammelas
Phylum Chordata / Subphylum Vertebrata / Class Osteichthyes

Midwater eelpout, about 7 cm long. From outer bay, 800 m depth. Deep sea laboratory, Monterey Bay Aquarium.

Most fishes in this family, Zoarcidae, inhabit continental shelves and slopes. They are believed to have evolved in the arctic some fifty million years ago, and they have spread as far as the Antarctic Ocean (M. Anderson 1994). Interestingly, antifreeze proteins are present in the body fluids of many of these cold-water fishes, suggesting that climatic stress during a later ice age favored individuals that developed such proteins before this fish family diversified and spread (Shears et al. 1993).

Adult *Melanostigma*, as the generic name (meaning "dark mark") suggests, are black or dark brown—this translucent, bluish one may be a juvenile. The fish have a characteristically blunt head, loose skin, and no scales. Eggs are brooded, and the young leave their mother some four months after fertilization.

Melanostigma is a pelagic species that lives in the deep waters of the OMZ, where it reduces its metabolic rate to conform to the markedly limited oxygen levels there. It does not migrate vertically. It tolerates the increased oxygen concentration of surface waters, however, and as a result has become a "laboratory rat" for physiological experiments.

This *Melanostigma* swims about in a darkened, chilled aquarium. When it is illuminated, it curls up in a ball, as seen in the photo, and hangs motionless in the water, later swimming about again. In the outer bay, when a deep-diving vehicle approaches and shines a light on the swimming *Melanostigma*, the little fish does exactly the same thing. To a predator approaching in dim midwater light, a curled-up eelpout may appear too large to swallow; or perhaps it resembles a small jellyfish with low nutritional value and potent nematocysts. One thing is certain: it no longer looks like a fish. Bruce Robison (1995), who has spent countless hours observing midwater creatures, has seen hake strike at a fleeing fish while ignoring a curled-up eelpout nearby.

5 Subtidal Reefs

As we have seen, intertidal rocks show striking zonation. Subtidally, below the physical stresses associated with tidal rhythms, zonation is not as sharply demarcated. At six meters' depth the rocks may be covered with dense algae, mostly foliose reds. With abundant light, these seaweeds evidently can outcompete sessile invertebrate animals for space on the substratum. Nevertheless, in rocky crevices and on shaded north-facing vertical surfaces you will find solitary corals, clones of strawberry anemones, and many other invertebrates. At a depth of some ten meters, algal cover thins in dimmer light and invertebrates begin to dominate.

Much giant kelp arises from holdfasts at depths of ten to twenty meters, and the substratum there often is shaded beneath its canopy. With natural lighting alone, it's like being outdoors on a moonlit night.

Here, Lovell surveys a favorite subject, walls of colorful *Corynactis* anemones. Libby took this photo, illuminating her subjects with an electronic flash unit. Since water rapidly filters out the red end of the visual spectrum, both her camera lens and her flash unit must be held close to the anemones to record their brilliant, warm color. At this depth, background subjects beyond the strobe's reach, like the kelp here, appear blue-green, which is the wavelength of the more deeply penetrating light. Without direct illumination, the underwater world's "true" colors go unrecognized.

One of the diver-author-photographers at work, Carmel Bay, 12 m deep.

Life on a ledge. Carmel Bay, 18 m deep.

This black and yellow rockfish, *Sebastes chrysomelas*, regards us with a wary eye. A bottom-dwelling territorial fish, it may live on this ledge for years, driving away intruders but ever ready to pounce on juvenile rockfish. In contrast, the other animals here, such as the white-spotted rose anemone *Urticina lofotensis*, the solitary corals, the hydroids, the bryozoans, and the sponges, are all sessile suspension feeders. They compete for living space, not for food; after all, the ocean's surge will bring suspended food particles to them. Small patches of coralline reds are the only algae here.

Above right: Invertebrate diversity of the open coast. "The Pinnacles," 26 m deep.

We have reached the Pinnacles—an exposed site renowned for its rich invertebrate fauna. Even at this depth long-period swells transmit a powerful to-and-fro surge to the bottom, which the invertebrates seem to love. By examining this image with magnification, we make out five bryozoan species, four cnidarians, two colonial tunicates, two echinoderms, two sponges, one polychaete worm, and one encrusting coralline red alga; and surely there are tiny creatures crawling about, hidden in the thicket. All this in an area less than a meter across. Extraordinary! And not a square centimeter of bare rock to be seen—there are just too many larvae seeking a place to land, and lots of planktonic food to nourish this faunal array.

Exploring such a subtidal reef is not like seeking wildflowers in a field, however. Many readers may not be fully aware of the demands of diving, particularly the effects of increased pressure. At this depth of 26 m, for example, your wet suit becomes compressed to less than half its normal thickness; insulation is lost, and the cold can be numbing. With a compressed wet suit, too, your own volume is reduced, you become relatively heavy, and you must add air to your buoyancy compensator to preserve neutral buoyancy. Not only your wet suit, but even the air you breathe is compressed; it is now less than half its surface volume, which means you are using it at more than twice the rate you would in shallow water. So you must keep checking your pressure gauge to determine how much air remains in your SCUBA tank—the last thing you want is to run out of air! Finally, since the body absorbs nitrogen rapidly at this depth and pressure, one must be aware that, to avoid bends, prolonging a visit in this paradise beyond some twenty minutes may require a decompression stop on the way back to the surface. Fortunately, with experience all these calculations become almost automatic, and dive computers help. But a mere twenty minutes? If only we could repeal the laws of physics and the limits of our physiology. . . .

Cystoseira osmundacea
Phylum Heterokontophyta / Class Phaeophyceae

Top right: Bead kelp, 1 m high. Carmel Bay, 15 m deep.

This brown alga may live for many years. Each thallus consists of a holdfast and a short, tough stipe that bears flattened blades. From the tips of the blades the alga each year sends slender reproductive fronds, buoyed by small floats, to the sea surface. This photo, taken in January, shows the alga swept horizontally by ocean surge, and reproductive fronds have just begun to appear. On the rocks surrounding *Cystoseira* are both upright and crustose coralline red algae.

Cystoseira thalli, which are diploid, may be male, female, or hermaphroditic. Annually, the reproductive fronds meiotically produce and then release eggs and/or biflagellate sperm. Fertilization and germination lead directly to another diploid thallus. Thus there are no sporophyte or gametophyte phases in the alga's life cycle; in contrast to giant kelp—and indeed most algae—*Cystoseira* has a life cycle that is animal-like.

Bottom right: Female reproductive frond of *Cystoseira,* about 18 cm, and an epiphyte. Carmel Bay.

This reproductive frond has reached the surface and is releasing eggs. It was photographed in August, when these fronds create their own tangled canopy. During winter it will be ripped away by storms and thrust ashore.

Note the small saccular brown alga growing amid *Cystoseira*'s developing eggs. This is an epiphyte, *Coilodesme californica*, which grows only on the reproductive fronds of *Cystoseira.* This sporophyte phase, beginning as a little sac, may attain a length of 30 cm. Presumably the gametophyte phase of *Coilodesme*'s life cycle is microscopic and carries the alga through the fall and winter when *Cystoseira*'s reproductive fronds are absent.

Fauchea laciniata

Phylum Rhodophyta / Class Florideophyceae

Red alga, female gametophyte, about 3 cm long. Carmel Bay, 23 m deep.

Underwater, this alga's blades appear a striking iridescent blue. We have found it at moderate depths (12–25 m) on vertical, shaded walls that appear studded with tiny blue swatches of the alga. Since mostly blue-green light penetrates to such depths, and any light at all is in short supply there, one would expect these red seaweeds in such circumstances to absorb blue light (as a red pigment does), not reflect it (as, of course, a blue pigment does). Why does the alga pose this chromatic paradox?

The paradox is resolved when one discovers that the alga's blue color is not due to a reflective pigment at all but rather is a "structural color"—iridescence results from the interference of light waves reflected from very thin, actually colorless layers in the alga's cell walls. In fact, when these seaweeds are taken out of the water, the blue color is lost within minutes as the cell walls dry and stop scattering light. The seaweed then appears red from its accessory photosynthetic pigment phycoerythrin, which was there all along. The red phycoerythrin is very efficient at absorbing deeply penetrating blue-green light.

The phases of *Fauchea*'s life cycle are isomorphic: that is, the gametophytes and sporophytes look alike. This is a female gametophyte, identified by the little excrescences over the blade. Sperm fertilize the female's eggs at these bulging sites. There, the resultant zygotes undergo complex developments through the next life cycle phase of red algae—the carposporophyte—to produce carpospores. Following their release, carpospores settle and develop into sporophytes (see Fig. 3).

Male gametophyte, immature female gametophyte, or sporophyte? About 3 cm long. Carmel Bay, 12 m deep.

It would take careful microscopic examination to identify this seaweed as a male gametophyte, sporophyte, or even an immature female gametophyte. At least we know that it's not a mature female gametophyte, since its surface is smooth. It is surrounded by coralline red algae.

Derbesia marina

Phylum Chlorophyta / Class Chlorophyceae

Top right: "Halicystis ovalis," Derbesia marina's gametophyte phase, 10 mm diameter. Open coast, 25 m deep.

Widely distributed in the northern hemisphere, the green algae *Derbesia marina* and "*Halicystis ovalis*" were first described from islands in the North Sea and named as distinct species. Then in 1938 Peter Kornmann, working at the German state biological laboratory in Helgoland, cultured the branching, filamentous green alga *Derbesia* and established that its spores gave rise to a vesicular gametophyte, the so-called *Halicystis* phase. Now, seen for what it really is, "*Halicystis*" has lost its old name, except in an informal way, as we have put together a more complete picture of *Derbesia marina*'s heteromorphic life cycle.

"*Halicystis*" lives only on crustose coralline algae, usually at moderate depths. Each "pea" is an enormous single cell with many nuclei, all haploid. The cells are hermaphroditic, producing both male and female gametes in yellowish tan fertile areas. One "pea" can successively release gametes at intervals of twenty-four hours. The biflagellate gametes mix in the sea; sperm surround the eggs, but only one sperm actually fuses with an egg. The resulting zygotes then sink and germinate to form new *Derbesia* sporophytes (R. Lee 1980).

Bottom right: Derbesia's sporophyte phase, on the gumboot chiton, *Cryptochiton stelleri,* 15 cm. Monterey Bay, 12 m deep.

Derbesia's sporophytes grow as branching, green filaments about 1 cm long, often on encrusting coralline algae or on rocks. Identifying them among so many similar filamentous green algae can be daunting, but sometimes the task is eased by this species' unusual habit of also growing on gumboot chitons, as seen in this photo. *Derbesia*, after complex development, meiotically produces haploid spores. The spores, each with a whorl of flagella, disperse and later settle and germinate into the haploid "*Halicystis*" gametophyte phase of *Derbesia* (Fig. 13).

Along our coast, the intertidal and the shallow subtidal rocks are covered with seaweeds, including many green algae such as *Ulva*, sea lettuce. We had never seen a green alga in deeper water until we found "*Halicystis*" at moderate depth. Our personal observations until that time matched the long-standing belief that greens are strictly shallow-water algae,

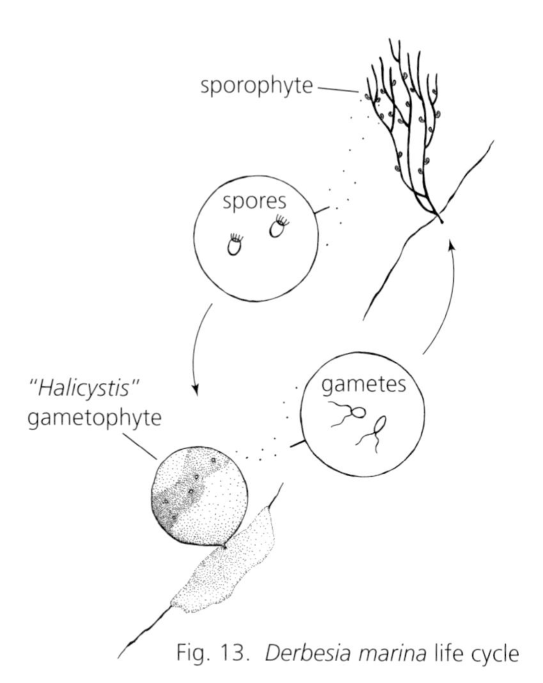

Fig. 13. *Derbesia marina* life cycle

poorly adapted to deep water. After all, the photosynthetic pigments of green algae (chlorophyll a and b, and some xanthophylls) mostly reflect, rather than absorb, green (hence their color), and deeper habitats in sunlit coastal waters have only dim green light. In contrast, the absorption pattern of brown algae's accessory photosynthetic pigments appears to adapt them to the light spectrum found at moderate depths, and that of reds' to still greater depths.

This familiar dogma, however, is wrong (Saffo 1987; Lobban and Harrison 1994). Red algae live along the California coast on shallow rocks as well as in deep water, and in Hawaii there is little difference in the relative numbers of green, brown, and red algae in either shallow or deep water. The sheer *quantity* of light, it turns out, is more important than its quality. Seaweeds adapt to the dim light far below the surface by increasing *all* their accessory photosynthetic pigments. At depth, the particular pigment complement of a seaweed is less important than its ability to absorb all the light possible where it lives. Also, the alga's shape and structure prove to be important, allowing dim light to be bounced about within the alga as though by mirrors or conducted through it as though by fiber optics (Saffo 1987). So *Halicystis* "belongs" where we found it, and this ecological appropriateness makes more sense when the alga's entire pigmentation and structure are taken into account.

Leucosolenia eleanor

Phylum Porifera / Class Calcarea

Sponge and bat star, *Asterina miniata,* field of view about 5 cm. Carmel Bay, 16 m deep.

Here we see the seastar *Asterina miniata* preying on the sponge *Leucosolenia.* Since sponges have extraordinary regenerative abilities, it is likely that multitudes of cells will rebuild and repair any damage inflicted by the bat star. Although a few adult sponges can slowly move about, obviously sponges cannot run away from potential predators. Over evolutionary time, however, sponges have developed formidable defensive spicules—needles and thorns—and they synthesize notoriously toxic compounds. Even so, many of them are preyed upon by nudibranchs, chitons, seastars, and even fish. The sponges' vast array of unusual chemical compounds has led chemists to explore the oceans seeking ones with antibiotic, insecticide, chemotherapeutic, or other activity useful to man.

Leucosolenia has a simple body structure featuring tubular chambers lined by "collar" cells (choanocytes), each with a long flagellum, that propel water through the sponge's chambers and toward each collar cell's base. There, tiny food particles, particularly bacteria, are efficiently caught by the collar cell's "hairs" and ingested. Water enters the chambers through minute pores, each in the center of a pore cell, and exits through the larger openings conspicuous in these photos. Pore cells can contract, effectively reducing the water flow. Obviously, a filter-feeding system can be efficient only if it is not plugged by the growth of bacteria and other organisms. The sponge's toxic compounds may prevent the growth of such fouling organisms on its surfaces. Nevertheless, a variety of protozoans and small worms and crustaceans find shelter in *Leucosolenia*. Finally, many of a sponge's cells are surprisingly plastic and can give rise to other cell types. For example, in a species of *Leucosolenia* (*L. complicata*), choanocytes give rise to sperm (Anakina and Korotkova 1989).

Some twenty years ago, *Leucosolenia* larvae readily settled and metamorphosed on panels submerged in Monterey Bay, and little sponges grew to 5 cm within three months (Bakus and Abbott 1980). That may not happen today, as *Leucosolenia* is much less common now.

A close look at *Leucosolenia*, field of view about 35 mm. Monterey Bay, 14 m deep.

Leucosolenia's simple tubular structure is well shown in this view; the tubes measure about 2 mm in diameter. Note the prominent spicules, seen through the sponge's translucent walls. Made of calcium carbonate, the spicules are linked together to form a hard skeleton that strengthens the body wall and may deter predation.

Sponges are indeed odd animals; lacking true tissues and organs, in many ways they are quite unlike other multicellular animals. And yet sponges share certain genes (HOM/Hox genes) with cnidarians and with more evolved bilaterally symmetrical animals that govern aspects of body design during embryonic development; these genes "tell" cells what part of the body they belong to. The finding supports the concept of a monophyletic origin of the kingdom Animalia (McGinnis and Kuziora 1994; Degnan et al. 1995).

There is growing evidence, moreover, that the ancestor of all multicellular animals may have been a pelagic organism, a primitive ball of cells with an epithelium consisting mainly of flagellated cells—yes, like choanocytes—that propel it! The sponges may have diverged early from these primitive organisms, and their current lifestyle evolved when an adult, benthic stage became established, the pelagic stage being retained in the larvae (Nielsen 1998).

What we can say is that sponges are an ancient, conservative lineage that has prospered and diversified in the sea probably as long as any animals at all have been there—surely an enviable record of ecological and evolutionary success. Their origin has been pushed back to 580 million years ago with the discovery in China of tiny fossilized sponges (Li, Chen, and Hua 1998). This is some 40–50 million years before the great Cambrian explosion that resulted in the diverse animal forms of today.

Tethya aurantia

Phylum Porifera / Class Demospongiae

Opposite: Orange puffball sponges, about 6 cm. Carmel Bay, 18 m deep.

Sponges rarely can be identified to species, and sometimes not even to genus, by casual inspection alone. Serious taxonomists identify sponges by their skeletal spicules and other internal details that often require microscopic examination of tissues. But these orange puffballs, surrounded here by the white-spotted rose anemone (*Urticina lofotensis*), a bat star (*Asterina miniata*), orange cup corals (*Balanophyllia elegans*), and several colonial animals, are distinctive even from the outside. *Tethya* is one of many "horny" sponges (demosponges) having glass spicules and skeletal fibers of spongin, a structural protein.

Tethya reproduces both sexually and asexually. Its eggs are fertilized internally and give rise to swimming larvae. A single settled larva can grow into a small sponge, but often several larvae that have settled close together will fuse, reorganize, and grow into one mixed "chimeric" sponge. *Tethya* also clones by budding from its surface or from rootlike stolons.

Regeneration experiments performed by H. V. Wilson more than a century ago (1891) cast light on the biology of sponges, particularly self/non-self recognition. If a sponge is pressed through fine cloth into seawater, the separated cells rapidly regroup in small clumps to form tiny new functional sponges. A protein-carbohydrate "aggregation factor" appears to mediate the response. The factor often is species-specific, but sponge cells of one species sometimes will aggregate with cells from another and form a chimera. This can happen with two species of *Tethya*. In such instances, perhaps the aggregation factor recognizes (or is fooled by) similarities between the separate species' cells.

Another way of exploring these phenomena is through experiments where pieces of the same or

different sponge species are grafted onto one another. Unfortunately, such grafting experiments, like reaggregations, cast both light and confusion on cell recognition in sponges. Grafts between two different *Tethya aurantia* sponges may fuse and function normally, or they may initially fuse but eventually reject each other. Sometimes grafts between one *Tethya* and another species of the genus will fuse, and other times not. So don't pursue this subject looking for simple generalities. It isn't even clear whether the same mechanism is involved in cell reaggregation and again in fusion/rejection in grafting experiments, similar though these phenomena superficially may seem to be (Bakus and Abbott 1980; Coombe, Ey, and Jenkins 1984; Simpson 1984).

In addition to its pronounced ability to regenerate damaged tissues, *Tethya* makes use of its spicules to isolate and shed parts of itself that become infected by bacteria. And these sponges have subtle ways to protect themselves from still other problems. *Tethya* manages to live in coastal waters that are highly polluted by powerful toxins. It may retain but detoxify some foreign substances with enzyme systems (dangerous if detoxification fails), and it can actually expel other toxins. Several organisms, a species of *Tethya* (*T. aurantium*) and the mussel *Mytilus,* as we have seen, make P-glycoprotein, a protein that binds to many toxins and transports them out of the animals (Kurelec and Pivčević 1992; Müller et al. 1996). This P-glycoprotein "pump" not only has been found in several invertebrates, but it sometimes manifests itself with devastating consequences during cancer chemotherapy in humans. At such times, multidrug resistance suddenly occurs in tumors (e.g., some leukemias) previously responsive to anticancer drugs. The newly resistant tumor cell-lines show reduced concentrations of the chemotherapeutic drugs and increased concentrations of P-glycoprotein. Just like sponges, the tumors have protected themselves from toxins with this molecule (Kartner, Riordon, and Ling 1983; Kuwazuru et al. 1990).

Stylaster californicus
(= Allopora californica)
Phylum Cnidaria / Class Hydrozoa

California hydrocoral, field of view about 60 cm. Carmel Bay, 18 m deep.

On the open coast, large pink and purple colonies of these beautiful animals adorn reefs and walls, extending from moderate depths to beyond recreational diving depths. In this view colonies are surrounded by a green sponge, clones of strawberry anemones (*Corynactis californica*), and purple sea urchins (*Strongylocentrotus purpuratus*).

These are hydrozoan corals, not anthozoan "true" corals. The distinction may sound like a detail, but in fact hydrozoans and anthozoans differ greatly both in their life cycles and in their anatomy. Hydrocorals have at least the merest hint of a hydrozoan medusoid stage (completely absent in anthozoans) in their life cycle, and the simple hydrozoan body cavity lacks the dividing walls or membranes (septa) present in anthozoan polyps.

Colonies of *Stylaster* have separate sexes. Where the colonies are clustered, sperm shed by the retained medusoids of male colonies fertilize practically 100 percent of the eggs in the medusoids of female colonies. Tiny, nonfeeding, planula larvae escape from their mothers in late October or November and establish new colonies, usually nearby (Ostarello 1976). Few new colonies survive their first year, however; many attach to unstable surfaces or are outcompeted for space by other sessile invertebrates or algae, and those on horizontal surfaces can be smothered by drifting sediments. For the lucky survivors, growth is slow. Large colonies such as these could be up to one hundred years old. It is distressing to find these venerable colonies sold in local curio shops. We can only hope that eventually there will be an end to the looting of these and other treasures from our waters.

Opposite top: Two colonies, view about 15 cm. Carmel Bay, 20 m deep.

The beautiful colors of these colonies come from chemically altered carotenoid pigments that are bonded to the hydrocoral's calcareous skeleton. The pigments make their way up the food chain, possibly from diatoms via copepods. In the pink colony, the polyps contain oxygenated carotenoids. In the purple colony, one of these compounds (astaxanthin) is joined to protein and bonded to the calcareous skeleton (Fox and Wilkie 1970).

A commensal polychaete worm, *Polydora alloporis,* establishes itself by boring twinned holes, conspicuous here, into *Stylaster*'s skeleton—the worm's exclusive habitat. Often the worm's two tentacles protrude from

the holes. Most *Stylaster* colonies, such as the pink ones in the preceding photo, also harbor obligate commensal barnacles, *Armatobalanus nefrens* (= *Balanus nefrens*), which live embedded in knoblike swellings that have a single larger hole. The larval *Armatobalanus* settles on the hydrocoral, which then grows around it (Ostarello 1973). When it feeds, the barnacle's feathery appendages rhythmically protrude from its lair to capture plankton.

Right: Close-up, field of view about 35 mm. Carmel Bay, 20 m deep.

The prominent tentacles seen in this view are those of nonfeeding polyps (dactylozooids) that protect the colony and capture food. Deep within the colony's small stellate pores lie single feeding polyps, each with four tiny tentacles surrounding its mouth.

Corynactis californica

Phylum Cnidaria / Class Anthozoa

Opposite: Strawberry anemones, each 2 cm wide. Monterey Bay, 8 m deep.

Corynactis is one of our reefs' most prominent and striking sessile animals. Large clones decorate vertical rocks in vivid colors. This common low intertidal and subtidal soft polyp is more closely related to corals than to the anemones it at first glance resembles. It has clubbed tentacles and many other features of true corals but lacks their calcium carbonate skeleton. Its nematocysts are unusually potent. Small as the polyp is, a brush with its tentacles can make sensitive skin tingle; and some predators of anemones, such as the nudibranch *Aeolidia papillosa* (discussed in chapter 2), avoid this one.

Two clones, field of view about 6 cm. Carmel Bay, 20 m deep.

These two groups of *Corynactis*, in shades of red, are excellent examples of cloning. *Corynactis* forms clones by fission, generating genetically identical clusters of polyps, all of the same sex and of one color. Although the rate varies, a clone can double the number and hence the surface claims of its polyps in two months (Chadwick and Adams 1991).

Aggression between *Corynactis* clones has not been documented, but clones often are separated from each other as in this photo, suggesting aggressive behavior of a sort that has been missed so far by investigators. In contrast, *Corynactis* has been observed to engage in marked and often lethal aggression against nearby anemones such as *Anthopleura elegantissima* and *Metridium senile*, and against the solitary corals *Balanophyllia elegans* and *Astrangia lajollaensis* (Chadwick 1987, 1991). Like reef-building corals, *Corynactis* fights and defends itself with nematocyst-laden filaments extruded from its mouth.

Gonads develop in the fall, and the animals spawn in December (Holts and Beauchamp 1993). Thereafter, planktonic planula larvae disperse, settle, meta-

Corynactis dividing, view about 8 cm. Carmel Bay, 15 m deep.

This *Corynactis* polyp has almost completed the slow act of pulling itself apart. The polyps can separate by longitudinal fission into two or even three equal parts. They also can reproduce by fragmenting parts of their foot. The topsnail *Calliostoma annulatum*, caught in our photo, is one of the few known predators of *Corynactis*.

morphose, and start new clones. Sexually derived polyps are, of course, genetically different from one another, and this variation is often revealed by their differing colors.

Although carotenoid pigments occur in the tissues, *Corynactis*'s colors appear to be due principally to a water-soluble, bilelike pigment (West 1979). Since water filters out the red end of the visual spectrum, most red organisms look black when viewed at depth with available sunlight. Under the same conditions, however, *Corynactis* may have a rosy glow, apparently because some of its pigments fluoresce in response to ultraviolet radiation (Wicksten 1989).

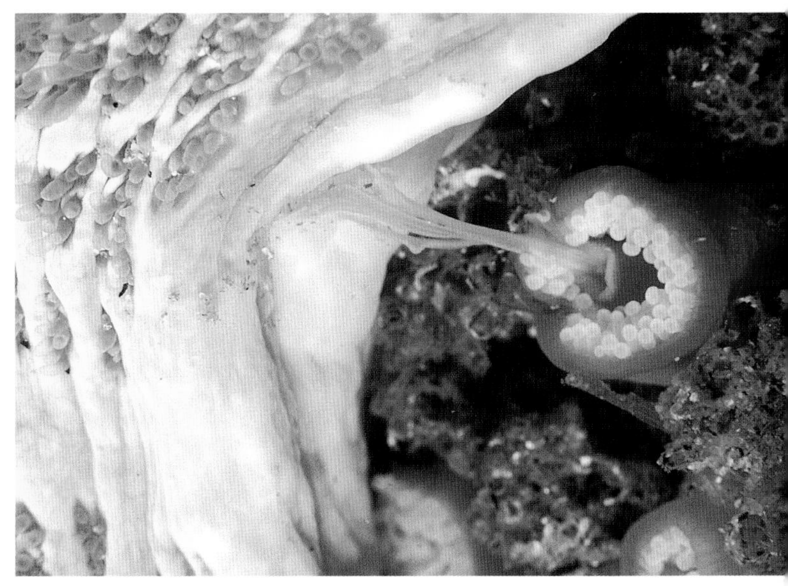

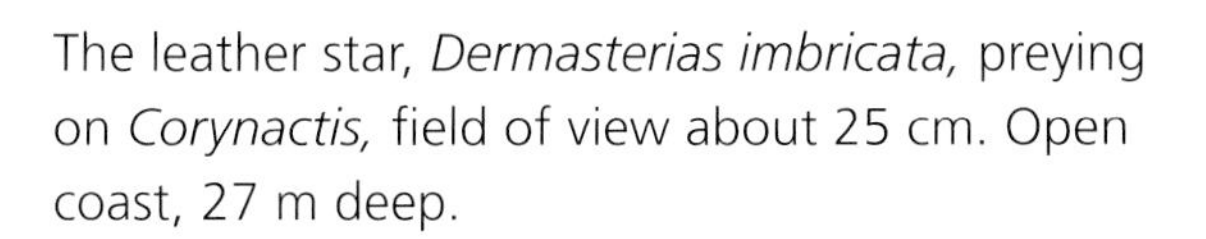

The leather star, *Dermasterias imbricata,* preying on *Corynactis,* field of view about 25 cm. Open coast, 27 m deep.

This leather star is hunched in its characteristic predatory embrace over *Corynactis.* It is another of the few predators known to attack this anemone; indeed, field studies have shown that *Corynactis* makes up a large part of its diet (Annett and Pierotti 1984). Initially, *Dermasterias* approaches *Corynactis* with its arms uplifted and retracts any tube feet that come in contact with it. *Corynactis*, unable to detach and float away, remains expanded. The seastar then assumes a domelike posture over its prey, minimizing contact with the nematocyst-laden tentacles, and gingerly swallows the polyps whole. Rapid cloning enables *Corynactis* to maintain itself in large numbers despite these assaults.

Corynactis defending itself.

Under attack, *Corynactis* has extruded nematocyst-laden filaments from its body cavity to repel *Dermasterias*. Note that the filaments adhere to the leather star's skin; they will remain attached and continue to sting *Dermasterias* as it retreats. *Dermasterias* may approach several groups of *Corynactis* polyps and then move away after making contact, apparently in deference to their defense.

Dermasterias imbricata

Phylum Echinodermata / Class Asteroidea

Opposite: Leather star, 15 cm across. Open coast, 27 m deep.

Here *Dermasterias* is surrounded by and even touching the orange cup coral, *Balanophyllia elegans*, but it is not hunched over the corals and may not be feeding on them. Annett and Pierotti (1984) observed encounters between the seastar and the cup coral. On every occasion the seastar lifted its arms and moved away. A likely explanation is that *Balanophyllia*'s nematocysts are even larger and presumably more potent than those of *Corynactis.*

As we have seen, *Dermasterias* preys principally on *Corynactis,* swallowing polyps whole, one at a time. Often the seastar has as many as ten polyps in its stomach, despite their nematocysts. In laboratory experiments *Dermasterias* takes small aggregating and plumose anemones (*Anthopleura elegantissima* and *Metridium senile*) in marked preference to *Corynactis.* Large aggregating and plumose anemones, however, remain expanded while under attack, and their tentacles can reach *Dermasterias*'s vulnerable dorsal surface, where their sting causes the seastar to back off (Annett and Pierotti 1984). These observations may help explain why clones of the relatively small aggregating anemones, so common in the intertidal zone, do not spread subtidally: they are too vulnerable to attack by *Dermasterias.* In contrast, giant green anemones and large solitary *Anthopleura* anemones thrive in subtidal surge channels, and large *Urticina* anemones are common subtidally. *Dermasterias* apparently "knows better" than to attack them.

In contrast to many seastars, *Dermasterias* has a soft, slimy feel to the touch. Its skeletal plates are embedded in the body wall, and it does not have pincer-like pedicellariae on its surface.

We already have presented examples of self and non-self recognition in a sponge, a cnidarian, and a tunicate. Echinoderms share this fundamental animal property. In an experiment, a patch of skin was removed from a *Dermasterias* seastar and grafted to another place on the same individual (autograft); similar pieces were also grafted to another individual (allograft). In autografting, healing followed promptly and the graft "took." In allografting, after a long delay (more than two hundred days) the graft was rejected. Rejected grafts showed infiltration by defensive cells and eventual tissue death. Should allografting be repeated three times in the same seastar, rejection time is greatly accelerated, occurring in only about eight days (Karp and Hildemann 1976). In other words, the leather star recognizes self and non-self. The response appears to be mediated by immune system cells that have memory—a system that seems analogous to that in vertebrates.

Additional evidence of an echinoderm's immunologic response having memory: in the laboratory, upon *Dermasterias*'s initial encounter with *Corynactis,* the anemone's toxins appear to temporarily paralyze *Dermasterias,* and on average it takes the seastar over forty-three hours to ingest its prey. However, in an encounter one week later, ingestion takes on average only ninety minutes! Interestingly, if one month elapses without contact between the two protagonists, the seastar loses its protective edge and returns to its more vulnerable condition (Willard 1981). These observations help to explain the seeming contradiction implied by one instance when the seastar is repelled by *Corynactis*'s defensive weapons and another instance when its stomach is full of the feisty little anemones.

Epiactis prolifera
Phylum Cnidaria / Class Anthozoa

Proliferating anemone, 3 cm across, on the brown alga *Cystoseira osmundacea.* Carmel Bay, 10 m deep.

This little anemone, perched on a blade of kelp, functions first as a female and later as both male and female. Oddly for an anemone, it broods its young, which attach to the spreading base of their parent's column. Intertidally *Epiactis* occurs on rocks, but subtidally we usually have found it growing on the brown alga *Cystoseira.* There the anemones are generally olive green, matching the alga; but, feeding on plankton, they do not gain their color from the kelp. Their color camouflage is probably adaptive, for they are preyed upon by fish (Yoshiyama et al. 1996). But the leather seastar, *Dermasterias imbricata* (surely nonvisual), also preys on these anemones and often lurks nearby. *Cystoseira* may provide a refuge for *Epiactis*, since seastars are not nimble climbers. Intertidally, *Epiactis* often moves into nooks and crannies where many predators cannot fit.

Side view of polyps, about 25 mm, attached above and below a blade of *Cystoseira*. Carmel Bay, 8 m deep.

Here a parent broods its tiny young; it may be their father as well as their mother. Younger *Epiactis* anemones are all female; older ones develop testes and become functional males as well as persisting as females. Various mechanisms inhibit self-fertilization in many hermaphroditic animals, but in this species self-fertilization occurs with regularity. Studies of the anemones' enzymes and DNA have shown that all the brooded young are either genetically identical or very similar to their mother, suggesting self-fertilization or extreme inbreeding (Edmands and Potts 1997).

The mother releases eggs from her mouth that have already been fertilized within her body cavity. Cilia then move the embryos across the oral disc and down her column until mucus and specialized nematocysts anchor the babies in place around her skirt. Avoiding the perils of the plankton, the young remain on their mother's base. They probably derive their nourishment from yolk reserves until they are able to scavenge debris from their mother's own feeding. At a length of about 4 mm, when the juveniles are strong enough to break the adhesive bonds to their parent, they start to crawl away to begin an independent life. Only about 20 percent survive the transition to adulthood, and small ones separated from their mothers do not survive (Dunn 1977).

Left: Proliferating anemone under attack by the nudibranch *Aeolidia papillosa*, 30 mm. Carmel Bay, 8 m deep.

Here a nudibranch predator has invaded the anemone's refuge on *Cystoseira* to launch an attack. *Epiactis* has withdrawn its tentacles and is closing up. More effectively, it can, if forced, detach itself and float away, an escape strategy employed by many anemones but probably a risky one. Notice how *Aeolidia*'s waving cerata closely resemble the tentacles of its prey. The advantage of this crypsis, if any, isn't clear, as most fishes reject eolid nudibranchs.

Opposite: Epiactis, about 25 mm, on rocky substratum, surrounded by articulated coralline red algae. Carmel Bay, 10 m deep.

This photo documents the unusual occurrence in Carmel Bay of *Epiactis* on subtidal rock rather than on *Cystoseira.* The polyp matches the color of the articulated coralline red algae surrounding it, in striking contrast to its color when it perches on kelp.

Paracyathus stearnsii

Phylum Cnidaria / Class Anthozoa

Opposite: Brown cup coral, 20 mm. Monterey Bay, 15 m deep.

In this view *Paracyathus*, a dime-sized solitary cup coral, is surrounded by a clone of smaller cup corals, *Astrangia lajollaensis*. These are among the few "true" coral species found along the California coast. The polyps live in calcium carbonate cups, their tentacles are arranged in multiples of six, and their body cavity is divided by septa. They do not harbor symbiotic algae, however, and they do not build anything like the huge coral reefs that are such spectacular features of the tropics.

The sexes are separate. *Paracyathus* releases large numbers of sperm or small eggs that are fertilized in the sea and develop into long-lived pelagic larvae. Successful establishment of new corals occurs infrequently, but the species lives well over forty years, barring physical damage (Fadlallah 1981; John Pearse, pers. comm.).

This coral's nematocyst clusters stand out as prominent white beads on the translucent tentacles. Nematocysts are intracellular organelles, the largest and most complex known, that act as microscopic harpoons. They usually consist of hollow, coiled tubules, armed with barbs, that evert explosively to penetrate prey and inject a neurotoxic venom of phenol and proteins. Nematocysts are generally regarded as independent effectors—that is, they are not discharged by the animal's nerve net, but rather by stimuli that are applied directly to them. However, a hungry anemone's nematocysts are more trigger-happy than those of a satiated one.

To fire, nematocysts require direct chemical *and* mechanical stimulation. For example, certain chemicals of the prey (including one released by wounded organisms, glutathione) are recognized by sensory hairs on a nematocyst's supporting cells. And a sensory hair on the nematocyst itself responds to mechanical stimulation, for example to the vibrations of a swimming crustacean. It is suggested that nematocysts fire following a cascade of events: (1) The supporting cells recognize a prey's chemicals, calcium ions rush into the supporting cells, and these cells release nitric oxide, a chemical messenger that has been discovered only recently (Salleo et al. 1996). (2) Nitric oxide reaches the resting nematocyst, increasing its excitability, and the nematocyst fires when mechanically stimulated. The whole event takes about three milliseconds, one of the fastest cellular processes known (Holstein and Tardent 1984; G. Watson and Hessinger 1989; Salleo et al. 1996). The nematocyst's sensory hairs are quite similar to those in our own inner ears. Nitric oxide modulates arterial blood vessel tone in mammals by relaxing smooth muscle cells, and it acts as a signaling molecule in the nervous system, among other actions. In 1994 it was named "molecule of the year" by *Science* magazine, and in 1998 three U.S. researchers who elucidated its physiology were awarded the Nobel Prize.

When *Paracyathus* and *Astrangia* polyps touch each other naturally, as in this photo, no aggression occurs between them, nor do they fight with each other in the laboratory. But when *Paracyathus* meets the orange cup coral, *Balanophyllia elegans,* or the strawberry anemone, *Corynactis californica*, aggressive interactions greatly influence the polyps' pattern of distribution. For example, *Corynactis* will move away from and avoid contact with adjacent *Paracyathus* corals, but *Corynactis* injures or kills adjacent *Astrangia* and *Balanophyllia* corals (Chadwick 1991). So there appears to be a competitive hierarchy of dominance between these species, one that influences their respective distribution—and *Paracyathus*, with the biggest nematocysts, is the bully.

Paracyathus, about 15 mm, under attack by the nudibranch *Hermissenda crassicornis.* Monterey Bay, 20 m deep.

Every bully, though, has its come-uppance! This coral, under attack by *Hermissenda,* has withdrawn its tentacles in retreat—revealing, in the process, the coral's calcium carbonate cup and septa.

Eudistylia polymorpha
Phylum Annelida / Class Polychaeta

Feather-duster worms, each about 20 mm across. Carmel Bay, 15 m deep.

These sabellid worms (family Sabellidae), here surrounded by coralline algae, live in tough parchment-like tubes. Each extends a plume of feathery, ciliated gills, maroon, orange, or brown. The worm uses the gill plumes to obtain oxygen and as a food-gathering net to filter microscopic particles from the water. *Eudistylia* has two giant nerve trunks that run the length of the body along the ventral (lower) midline, innervating the longitudinal muscles of whole-worm contraction. Eyespots on the plumes are sensitive to light. When the worm detects an "ominous" shadow or untoward vibrations in the water (rather than merely the sway of the ocean's surge), it instantaneously pops back into its tube, as one has done on the left in this photo. The end of its tube folds shut after the withdrawal, providing additional protection from predators. Should a fish be quick enough to bite off its head and plumes, *Eudistylia* can regenerate the missing parts. In fact, one related species of tube worm regenerated its head and tentacles thirty times in succession, though there was a progressive reduction in the worm's size (Carefoot 1977).

Eudistylia gill plumes, 20 mm across. Carmel Bay, 15 m deep.

The gills in the plume have complex sorting tracts of beating cilia that reject heavier inedible particles and move trapped bits of food down channels toward the mouth in the middle of the plume. This photo shows the rows of minute, photoreceptive black eyespots on the stiff supporting rods of some of the gill plumes. These are compound eyes that suggest those of arthropods. In these worms, each tapering cone within the compound eye has a powerful lens, a narrow visual field, and a single photoreceptor cell at the bottom. With several eyes along each gill plume, one worm may have as many as 240 compound eyes, each with 50 visual cones (ommatidia), which adds up to 12,000 overlapping visual fields. The resolution of compound eyes is poor, but the many very narrow visual fields, taken together, are acutely sensitive to motion. It's a system designed to detect the shadows of predators, without being triggered by particles in the water or very small planktonic organisms (Nilsson 1994).

Despite this system, there must be other photic receptors at play as well. One species of sabellid worm, for example, still shows a shadow reflex after its tentacular compound eyes have been experimentally removed. And closely related tube-dwelling polychaetes (serpulids and spirorbids), though they completely lack tentacular compound eyes, respond robustly to shadows. Sabellids, for one, do have a backup visual system consisting of simple pigmented eyespots—not compound eyes—on their heads, which adds to their ability to protect themselves (Nilsson 1994).

Serpula vermicularis

Phylum Annelida / Class Polychaeta

Below: Serpulid worm, each crown about 4 mm across. Carmel Bay, 10 m deep.

In contrast to sabellid worms, serpulid polychaetes (family Serpulidae) live in calcareous tubes, as seen here, not parchment ones. The species occurs on subtidal rocks down to 100 m depth, but it is cosmopolitan, often present in harbors on wharfs and docks. The little gems must be seen quietly and with magnification to be appreciated. In the field, even a stealthy approach may provide only a fleeting glimpse of these skittish animals. A two-part gill tuft about 1 cm across provides the worms with oxygen and traps tiny food particles that beating cilia carry to the mouth. As with sabellids, in serpulids sexes are separate, both sexes spawn their gametes, fertilization is external, and larvae swim only briefly before settling and metamorphosing.

Serpula gill plumes and opercula, about 10 mm across. Carmel Bay, 10 m deep.

One of the serpulid's gill plumes is modified to form a funnel-shaped operculum that stoppers the tube when the worm has withdrawn. The worm can regenerate this apparatus if a fish bites it off. The beautiful red color of this plume is due almost entirely to astaxanthin, a pigment the worm synthesizes by oxidizing the carotenoid beta carotene, which it obtains from algal food. This pigment (also present in the crowns of the sabellid *Eudistylia*) may serve to screen tissues from harmful effects of light.

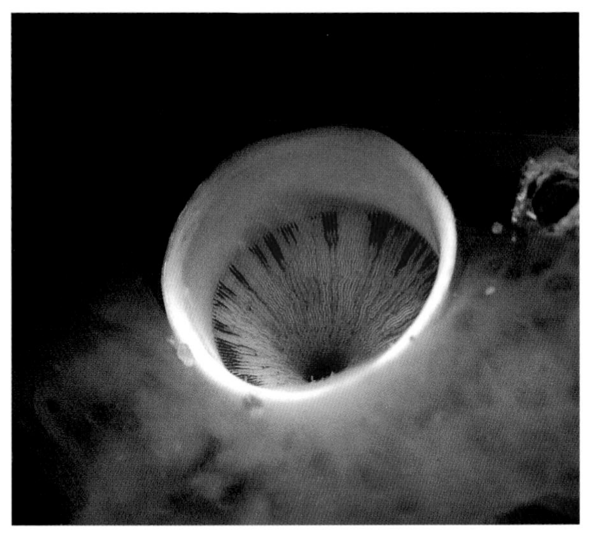

Right: Serpula calcareous tube and operculum, about 3 mm. Carmel Bay, 10 m deep.

We were too late this time! Photographing these tiny animals is a challenge. Although they seem not to mind the strobe's flashes, their sensitivity to shadows often defeats the photographer. Almost any vibration or shadow provokes a split-second contraction of the entire worm. But individuals living in an aquarium apparently become habituated to the disturbed conditions, for they no longer respond as quickly to passing shadows.

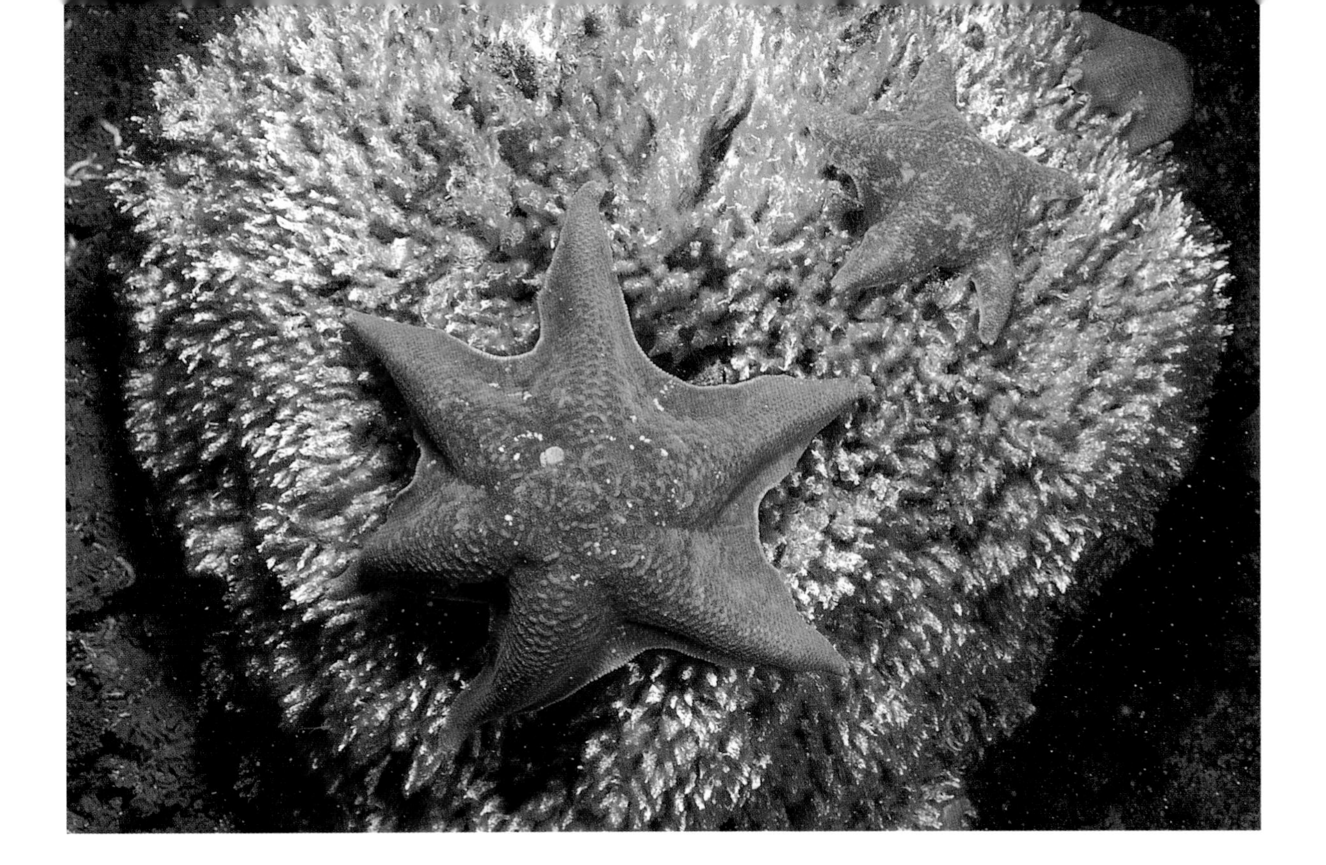

Salmacina tribranchiata
Phylum Annelida / Class Polychaeta

Fragile tube worms and bat stars, field of view about 22 cm across. Monterey Bay, 17 m deep.

From a distance we see the fuzzy orange clumps of these thin polychaete worms; but if we make the mistake of approaching too quickly, we suddenly face nothing but a tangle of hard, tiny white tubes. Like other serpulids, *Salmacina* retracts its plumes instantly when it detects unusual shadows or vibrations. Unlike, say, *Serpula*, however, it has no operculum. These minute worms are less than 5 mm long and live in clustered tubes less than 1 mm in diameter. The animal extends its orange gill plumes to feed on plankton. We have always found *Salmacina* low on vertical surfaces adjacent to silty sand; perhaps detritus makes up an important part of its diet.

The bat stars in the photo, *Asterina miniata*, seem to be feeding on the *Salmacina* worms. Although we did not remove the seastars to confirm this predation, apparently the seastars succeeded in inserting their stomachs into the tiny worms' tubes. Note that the adjacent worms have not reacted to this predation by withdrawing into their own tubes; *Salmacina* evidently lacks warning pheromones.

Close-up of *Salmacina tribranchiata*, field of view about 16 mm across. Monterey Bay, 17 m deep.

These worms are busily feeding, unconcerned by the nearby photographer. As you might guess from the appearance of these tightly packed tubes, they clone—in contrast to the related tube-dwelling *Serpula vermicularis* worms, which don't. *Salmacina* worms divide transversely in two, and then a gradual transformation of the two new ends regenerates two complete bodies. Finally, the new posterior worm breaks out through the wall of the tube and forms its own branch tube, while the new anterior one gets to keep the "parental" tube (Abbott and Reish 1980).

Petaloconchus montereyensis

Phylum Mollusca / Class Gastropoda

Opposite: Tube snails, field of view 17 mm. Monterey Bay, 17 m deep.

These sessile snails' tube-dwelling habit superficially resembles that of *Salmacina* worms: they both live in dense masses of tiny calcareous tubes. But inside the tubes live very different creatures—*Petaloconchus* is a snail that feeds with a mucus net, while *Salmacina* is a segmented worm that feeds with tentacles.

Petaloconchus tubes are 1–2 mm in diameter, far smaller than those of *Serpulorbis.* Unlike its larger relative, this snail bears an operculum, clearly visible in this photo, to seal its tube when it withdraws. Both snails extrude a mucus veil to trap tiny food particles, later to be reeled in and ingested. And they share the oddity of making some sperm that lack a nucleus and flagellum. The abnormal sperm disintegrate and may serve to nourish normal ones. Male snails discharge normal sperm in packets, to be caught in the mucus nets of the females (Abbott and Haderlie 1980).

Within her mantle cavity a female *Petaloconchus* lays up to a hundred eggs in individual capsules, and she can carry more than a dozen capsules. Just one larva usually develops and metamorphoses within each capsule, having subsisted on the many "nurse eggs," embryos that develop abnormally. For these snails, cannibalism is gender blind, occurring both in the sperm and in the embryos. Tiny juveniles emerge from their capsules, bypassing a planktonic larval stage. The liberation of crawling juvenile snails may explain the dense aggregations of the attached adults, whose tiny tubes can occur in concentrations of up to 100,000/m^2 (Hadfield and Iaea 1989).

Halosydna brevisetosa

Phylum Annelida / Class Polychaeta

Scale worm, 20 mm long. From shallow subtidal rocks, Carmel Bay.

This free-living polychaete reveals its anatomy better than its tube-dwelling relatives. The clumps of bristles along its sides, characteristic of polychaetes, act as cleats when the worm digs in its stumpy legs to crawl about. The worm's back is covered by eighteen pairs of large, overlapping, beautifully patterned scales. These scales shield the back of the worm and form a tunnel between the scales and the body surface through which beating cilia drive water to the gills. The worm casts off its scales if too roughly handled. The pink head bears tentacles and contains the "brain" and simple sensory organs such as the eyes, represented here by four dots. Each eye consists of a pigment cup and lens; the eyes probably perceive light intensity and direction, surely not an image.

We found this worm crawling among shallow subtidal algae, but the species often occurs as a guest in the tubes of other large polychaete worms and in moon snail shells occupied by hermit crabs, where it benefits from the crab's leftover scraps of food (MacGinitie and MacGinitie 1968). Free-living individuals generally are smaller than those living commensally, so perhaps "moving in" has advantages that free-livers simply miss out on, or maybe the commensal worms are just older. At times these animals are aggressive carnivores and fight with each other; they can evert their pharynx, which is armed with powerful jaws and four curved teeth, when the occasion demands.

Cucumaria miniata

Phylum Echinodermata / Class Holothuroidea

Opposite top: Sea cucumber feeding, tentacles 12 cm across. Carmel Bay, 13 m deep.

These soft-bodied sea cucumbers often lie hidden in cracks on vertical walls, but during plankton blooms they extend their sticky arms to feed. Each animal has ten arms that bear branching tentacles. When feeding, as in the photo, the fine tentacles on an arm roll up tightly, and the sea cucumber flexes its arm and thrusts it into its mouth. *Cucumaria* then "licks off" its armload of trapped food particles, the cleaned arm reextends, and its tentacles reexpand. Rhythmically, each arm in turn repeats this maneuver.

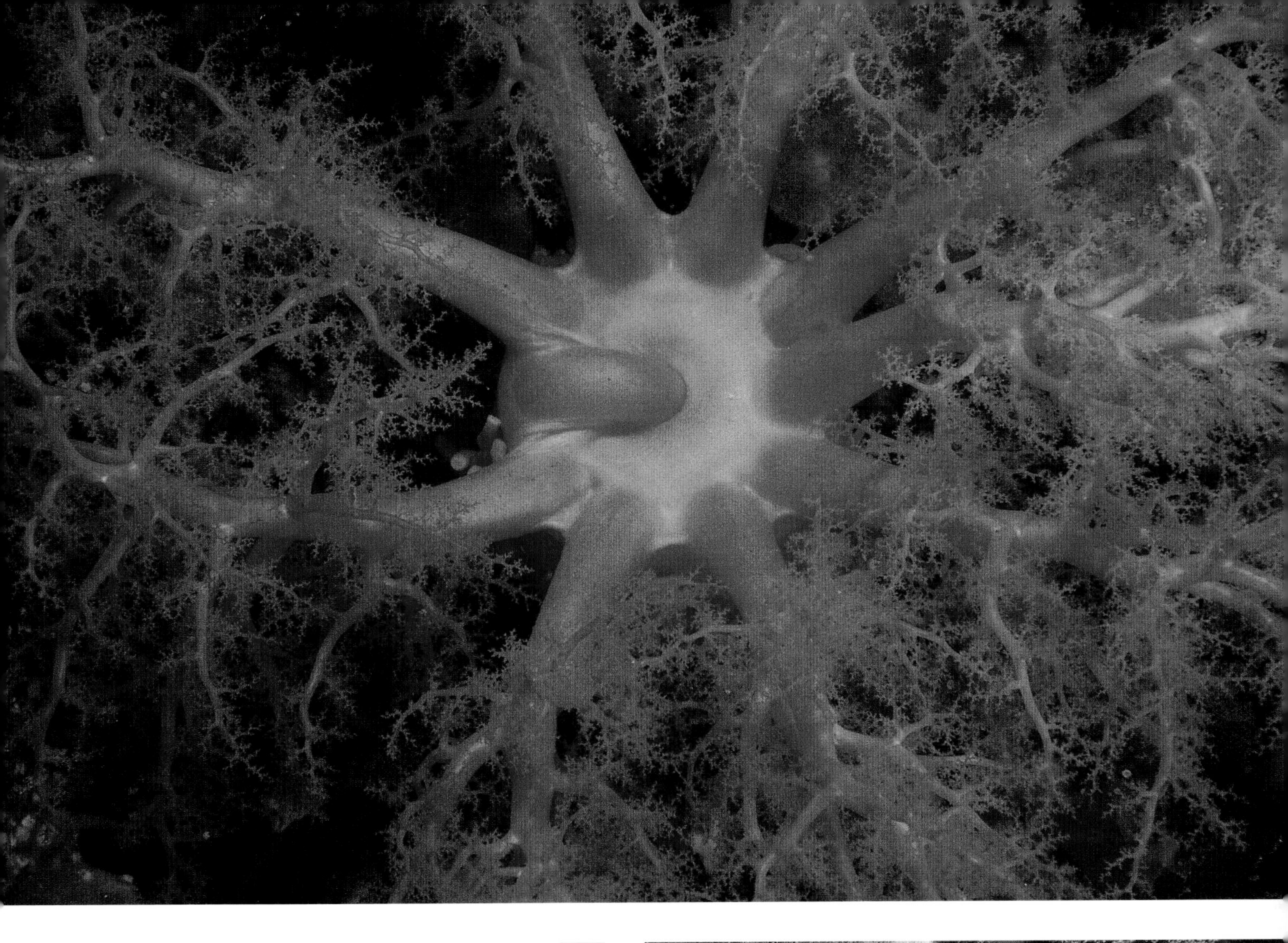

Right: Spawning *Cucumaria*, body extended 10 cm. Carmel Bay, 14 m deep.

Male and female *Cucumaria* spawn synchronously during spring phytoplankton blooms, though the larvae are nonfeeding. The spawning sea cucumbers stretch a third of their bodies out from the rocks, wave their tentacles, and sway from side to side as they release gametes. Bundles of buoyant eggs, together with sperm, are released in such numbers that they may form a slick on the sea surface. The great majority of the eggs—often nearly 100 percent—are soon fertilized (Sewell and Levitan 1992). This high success rate apparently is due to the proximity of the spawning animals to one another, the synchronous spawning of males and females, and probably the two-dimensional concentration of gametes at the surface. *Cucumaria* eggs are toxic to fish, but during their planktonic period of eight to thirteen days the larvae may be vulnerable to pelagic predators such as jellyfish.

Note the tube feet on the body of the extended animal. The suckers occur in five rows, consistent with the pentamerous symmetry of adult modern echinoderms.

If disturbed, the animal will rapidly retract into its crevice; sometimes *Cucumaria* will even messily eviscerate its gut, though the extruded material is not so toxic as that of certain tropical species (Carefoot 1977). The leather and sunflower seastars are among *Cucumaria*'s predators (Francour 1997).

This sea cucumber's bright color comes from very high concentrations of carotenoid pigments in its skin. Orange cup corals (*Balanophyllia elegans*), also colored by carotenoids, lie in the background. The pigments could have passed up the food chain from diatoms, which produce carotenoid pigments, to copepods, which eat diatoms and then fall prey to benthic suspension-feeding creatures like the sea cucumber and the cup coral.

Megabalanus californicus

Phylum Arthropoda / Class Crustacea

Opposite: Acorn barnacles, each about 5 mm across. On turban snail in kelp canopy, Monterey Bay.

Marine organisms often grow on other organisms that may live on still others. Here, for example, barnacle larvae have gregariously settled and metamorphosed on a turban snail and on each other. Now, as adult barnacles, they are busily extending their thoracic legs, fishing for plankton. The snail, in turn, is grazing on a blade of giant kelp encrusted with colonies of the bryozoan *Membranipora membranacea*. Note that the snail's shell is "fouled" on its spire not only by the barnacles but also by encrusting coralline algae. No doubt competition for limited firm substrata contributes to these intricate living arrangements, especially when they apparently do not harm the partners.

Young "nauplius" barnacle larvae feed while in the plankton; but for the older, nonfeeding "cyprid" larval stage that follows, the one goal in life is to find a suitable place to settle, preferably a hard surface that bears chemical traces of its own species. Clearly, with one barnacle nicely established on the turban snail's shell, others found it attractive to join the party. Now they are well positioned to mutually copulate and cross-fertilize each other.

Megabalanus is a particularly colorful barnacle that characteristically aggregates densely. The species occurs sporadically along the California coast and down into Mexican waters. In 1939 it turned up on buoys as far north as Humboldt Bay, perhaps reflecting conditions leading into the 1940–1941 El Niño event. In 1980 it was common subtidally about the Monterey Peninsula. In January 1988 off Hopkins Marine Station, an unusual, heavy settlement of *Megabalanus* occurred on kelp-dwelling turban snails, principally *Tegula brunnea* (Halliday 1988); the barnacles we illustrate may have been part of that event.

On another occasion, both *Megabalanus californicus* and a Panamanian species of *Megabalanus* were found on the hull of a fishing vessel in the Gulf of California, well out of either species' normal range (Newman and McConnaughey 1987). Shipping, as well as currents, undoubtedly augments larval dispersal of these animals that are benthic during their adult stage.

Didemnum carnulentum

Phylum Chordata / Subphylum Tunicata / Class Ascidiacea

Compound tunicate and blades of a red alga, about 40 mm across. Monterey Bay, 12 m deep.

Casual inspection of this filter-feeding colonial animal provides no clue that it is in the phylum Chordata. But when the tiny individual bodies (zooids), each only 1–2 mm long, reproduce sexually, their offspring disperse as the typical tunicate tadpole larvae with their distinctive chordate traits (as discussed in chapter 2 in connection with the tunicate *Botryllus*).

The zooids lie embedded in a common gelatinous tunic that is stiffened with microscopic, globular, calcareous spicules, rendering the entire colony yellow-white. Often, these thin encrusting colonies break up into many separate tiny colonial lobules. Evenly spaced on the colony's surface, each zooid has its own tiny oral intake pore; many zooids share the few larger excurrent pores. These tunicates sometimes are mistaken for sponges, and vice versa. Both have a gritty feel owing to their spicules, and the intake pores of *Didemnum* can contract to resemble the microscope intake pores of sponges.

Didemnum is hermaphroditic, as are almost all tunicates. During sexual reproduction the eggs are retained and fertilized internally, embryos are brooded, and tadpole larvae are released. With settlement, the larvae metamorphose into tiny zooids, losing their notochord and hollow nerve cord and developing a large perforated pharynx that filters particles from the water. The colonies grow quite big by replication of the minute zooids. The zooids' traits that reliably distinguish species or even genera are so minute and difficult to ascertain—the number of pharyngeal gill-rows, details of the sperm duct—that even specialists hesitate to tackle the taxonomy of these diverse and abundant animals. Probably several species are lumped together by neglect as "*Didemnum carnulentum*," awaiting a much-needed review of the entire family Didemnidae by someone dedicated (or maybe foolish!) enough to try.

Clavelina huntsmani

Phylum Chordata / Subphylum Tunicata / Class Ascidiacea

Opposite left: light-bulb tunicates, each 20 mm long. Monterey Bay, 15 m deep.

These clumps (actually colonies) of *Clavelina* appear in the spring and summer when upwelling of nutrient-rich water supports phytoplankton blooms. The zooids in this bouquetlike colony all are the budded progeny of a single settled tadpole larva; they are linked physiologically by blood vessels in the stolons

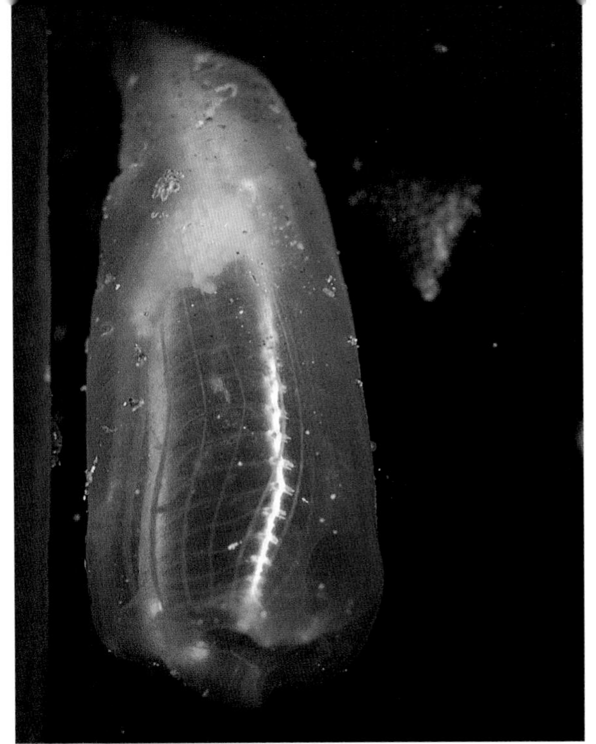

by which they spread. In the fall, as these zooids regress, tiny overwintering bodies will form at the tips of stolons that have grown out onto the substratum around the bouquet. New zooids will grow from these vesicles the next spring. Thus, while the zooids are "annuals," the colony may be quite old, restored to vigor many times.

The intake and excurrent pores can be seen at the tip of each zooid, and the transparent tunic reveals most of each zooid's internal structures. A glandular groove, called the endostyle, runs along the bottom of the long pharynx. Cells beside this gutter (part of the glowing "filament" of the "light bulb") extract and concentrate iodine from the sea and make a compound akin to thyroxin, a thyroid hormone in vertebrates. So the ascidian's larval notochord and other famous evidence quite apart, both the chemistry and anatomy of its endostyle suggest an affinity between the animal depicted here and vertebrates, including us.

In tunicates the endostyle secretes mucus sheets that slide over the inner wall of the pharynx—the transparent cylindrical structure in the photo—occluding patches of gill slits as they go. Water drawn in through the intake pore by a cilia-driven current is filtered through this mucus, which traps minute food particles. The filtered water then passes through the gill slits as a respiratory current and finally returns to the sea via the excurrent pore. Merely driving water through the cilia-laden gills for both respiration and feeding is widespread among marine filter feeders, from molluscs and annelids to phoronids and tunicates; but using a perforated pharynx to accomplish this job suggests the tunicate's evolutionary relations with vertebrates—a notion reinforced by the many other tunicate-chordate parallels we have noted.

Above right: A brooding *Clavelina* zooid, 20 mm tall. Monterey Bay, 15 m deep.

Although almost all simple tunicates discharge their gametes into the sea, colonial tunicates brood. Thus when *Clavelina*'s zooids release sperm to the sea, fertilization and brooding of the embryos occurs within other zooids. The cluster of eggs being brooded here lies within the atrial cavity of the zooid at the base of the pharynx. A little later in the embryos' development, one can open such zooids, shake out the developing larvae, and, with the stress of hypotonic seawater, induce metamorphosis into a newly formed zooid (Abbott and Newberry 1980).

Tonicella lineata

Phylum Mollusca / Class Polyplacophora

Opposite: Lined chiton, 15 mm long. Carmel Bay, 15 m deep.

Tonicella lives on crustose coralline red algae, its major food and the source of its own pigment. In this photo the well-camouflaged chiton is surrounded by partially closed *Corynactis* anemones, spirorbid worms, encrusting bryozoans, and a foliose red alga. The small size and the purple lines on *Tonicella*'s fleshy skirt—its "girdle"—are characteristic of the subtidal animals in this species. The muscular foot and the mouth with its filelike rasping tongue, or radula, lie completely hidden by the girdle and a set of eight shells (the latter of which inspired an early description of these molluscs as "coat-of-mail shells"). The radula contains the iron mineral magnetite, which is much harder than the algal crust's calcium carbonate that it grinds against. The mineral may be useful also as a navigational aid and in homing behavior.

Tonicella's disruptive color pattern, together with its location on crustose corallines, may help protect it from visual predators, but what about nonvisual predators? Subtidally in Monterey Bay, predatory seastars such as *Pisaster brevispinus* and *P. giganteus* have been observed to crawl near or even on *Tonicella* without taking the chiton. And in aquaria, *Tonicella* is rarely taken by predatory seastars or snails unless it is turned on its back, at which time it is readily consumed; the same is not true of two other species of intertidal chitons (Seiff 1975). It isn't clear why *Tonicella* is less vulnerable than other chitons; no chemical repellent has been demonstrated.

In addition to their eight shell-plates, chitons have eight pairs of pedal retractor muscles and many lateral gills in a groove between the foot and the overlying girdle of the body wall. Thus the morphology of chitons vaguely suggests the segmentation of annelids and arthropods. It remains highly debatable, though, whether these resemblances reflect underlying shared features that predate the historical separation of the mollusc and annelid lineages from a segmented common ancestor or whether they represent the chiton's own adaptations to clinging to surf-swept rocks.

Like most chitons, *Tonicella* is a broadcast spawner. In April, coincident with the spring phytoplankton bloom, females release small clusters of mucus-covered green eggs that are fertilized externally. The yolk-nourished larvae spend about a week in the plankton, their development arrested until they are triggered to settle selectively on crustose coralline red algae. They settle, as do abalone larvae, in response to a chemical that the alga releases (J. Barnes and Gonor 1973). For the abalone, the molecule that induces settling and metamorphosis is a protein or peptide related to gamma-aminobutyric acid, or GABA (Morse 1990). This same algal molecule may induce *Tonicella* settlement and metamorphosis. Interestingly, GABA is the mammalian brain's principal inhibitory neurotransmitter, and tranquilizers such as valium work by augmenting its action (Wickelgren 1999). How remarkable that GABA and its molecular relatives should accomplish such different tasks!

Why should the release of nonfeeding larvae be timed to a phytoplankton bloom? Himmelman (1975) speculates that since the peak abundance of carnivorous zooplankton occurs *after* phytoplankton blooms, *Tonicella,* with its short planktonic life, may be spared predation by spawning early with the phytoplankton itself.

Diodora aspera
Phylum Mollusca / Class Gastropoda

Rough keyhole limpet, 30 mm long.
Monterey Bay, 15 m deep.

This limpet, its shell covered by encrusting coralline algae, bryozoans, and polychaete worms, may be camouflaged from visual and tactile predators. But here *Diodora* has chemically detected the predatory seastar *Pisaster giganteus*, at the top of the photo, and displays its characteristic defense. As the seastar approaches, the limpet extends its soft mantle skin-fold to partially cover its shell from the bottom up, pokes its siphon through the "keyhole," and raises its shell from the substratum. Eventually the mantle will cover both *Diodora*'s foot and its shell. In the laboratory, *Diodora* shows this same defensive response to "seastar water" that has passed over a predatory seastar, an example of distance chemoreception. *Diodora* need not flee from *Pisaster;* its slippery mantle thwarts the seastar's tube feet. There may also be a chemical deterrent at work, for when the seastar's tube feet simply touch the limpet's enveloping mantle they often suddenly let go. In aquaria, this defense usually protects *Diodora* from *Pisaster giganteus* and *P. ochraceus* but not from the giant sunflower seastar, *Pycnopodia helianthoides*, which simply crawls over the limpet and swallows it whole (Margolin 1964).

In all shelled marine snails, heavily ciliated gills draw water into a "mantle cavity"—a deep pocket in the skin-fold next to the body—where respiratory water, fecal and urinary wastes, and gametes all collect and are released to the sea. In the presumed ancestral snail, the mantle cavity lay posteriorly, so these products poured out behind the animal. That's not

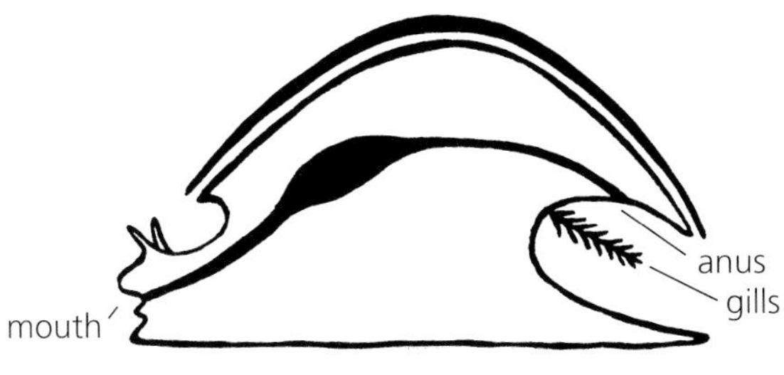

hypothetical ancestral mollusc

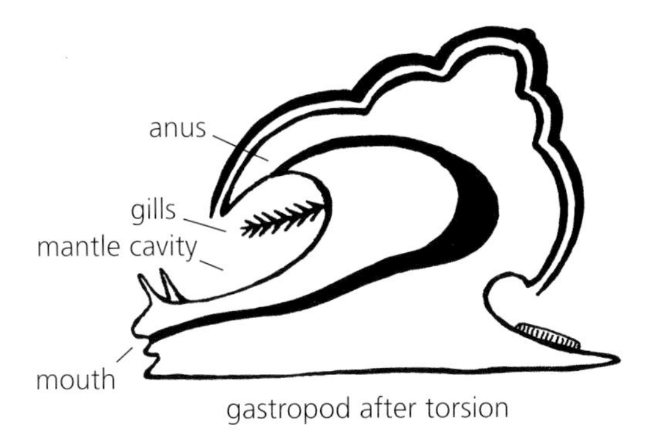

gastropod after torsion

Fig. 14. Gastropod torsion

the way it works today. Late in larval development, all contemporary snails undergo the strange phenomenon known as torsion, during which the shell and enclosed viscera rotate 180 degrees with respect to the head and foot (Fig. 14). This strange phenomenon introduces a potential sanitary problem: discharged waste products now could be released directly over the head. Contemporary snails have various solutions. The hole in keyhole limpets like *Diodora* (as well as the holes in abalone shells) reduces this handicap by allowing a one-way ventilating current of water to pass over the gills and carry metabolic and sexual products out the excurrent hole, which lies far from the head. Snails lacking holes in their shells have other "sanitary" adaptations, most notably ones that set the whole shell at an angle and tilt it, letting effluent water from the mantle cavity leave "over the right shoulder."

The poorly understood phenomenon of torsion, a radical realignment of body parts, is almost unprecedented in the animal kingdom. Though many theories have sought to explain its evolutionary advantage, as far as we know none has been supported experimentally. In one study, for example, torted and untorted abalone veligers were taken with equal frequency by predators (Brusca and Brusca 1990).

Curiously, when the apical shell opening of *Diodora* is blocked by the growth of an encrusting organism—or by an experimenter purposefully occluding it—the respiratory flow leaves the mantle cavity instead along the tips of gills that, in this crisis, extend especially far forward over the head. No apparent damage to the gills or mantle cavity results, but induced passive flow through the gills no longer can occur (Voltzow and Collin 1995).

Cypraea spadicea
Phylum Mollusca / Class Gastropoda

Chestnut cowry, 5 cm. Open coast, 25 m deep.

Of the usually warm-water cowry family only this species occurs on the Pacific Coast north of Baja California. We have found *Cypraea spadicea* at moderate depths on the open coast here, but in southern California we have seen it in shallow subtidal habitats. The cowry in this photo appears to be feeding on a yellow sponge; it also takes small anemones, the egg capsules of other snails, some compound tunicates, and occasionally bits of kelp. The nudibranch *Phidiana crassicornis;* a polychaete worm, *Serpula vermicularis;* a brittle star; a single strawberry anemone, *Corynactis californica;* and bryozoans add diversity to this view of a wonderfully rich reef. The encrusting coralline red alga and a sprig of foliose red alga play a subordinate role at a depth where invertebrates dominate.

Cypraea snails copulate, and females lay from 100 to 120 capsules, each with some 800 eggs. Until the eggs hatch in about three weeks, the mother guards the clutch by expanding her foot to cover the capsules.

People have collected cowries for their beauty since prehistoric times, but the living animal in its natural environment outshines any shell on a shelf. The mantle (partially raised in this photo) cleans and polishes the shell's shiny surface. It often covers the shell entirely, secreting additional shell material and preventing fouling by settling plankton. Cowry shells (and *Trivia* and *Olivella*, discussed below and in chapter 6) lack the dingy outer organic film, or periostracum, that coats most snails, so the glory of the shell itself radiates when the cowry retracts its mantle.

Trivia californiana

Phylum Mollusca / Class Gastropoda

Right: California trivia, 10 mm long. Carmel Bay, 17 m deep.

This beautiful little snail lives in Carmel Bay and along the open coast at moderate depths, usually on compound tunicates, its prey, as seen here. (The larger siphons of a solitary tunicate appear in the photo as well.) Except for the ribs on its shell, *Trivia* resembles a small cowry. This snail has only partially expanded its mantle, but it can raise it to completely cover its shell. *Trivia* feeds using its radula to scoop up both the tunic and the zooids of its prey, but it digests only the zooids. We have seen the snails deeply embedded within a compound tunicate, where they lay their egg capsules, from which the swimming larvae will escape.

In certain areas of the Norman and Breton French coast, species of *Trivia* (*T. arctica* and *T. monacha*) recently have shown almost a 100 percent incidence of genital abnormalities in which male parts, for example penises, occur in the snail's female genital system (Oehlmann, Stroben, and Fioroni 1993). These abnormalities are due to environmental pollution with tributyltin (TBT), a compound widely used as antifouling paint for shipbottoms. Closer to home, though far from snails, TBT may be indirectly responsible for sea otter deaths by compromising their immune systems. A sea otter that died at the harbor in northern Monterey Bay had a high concentration of TBT in its liver, and mussels—one of otters' prey items—taken in Monterey Harbor in 1988 contained the compound (Kannan et al. 1998). TBT use now is largely outlawed in this country, but, like DDT, residual amounts will persist for a long time, and it still is used elsewhere. As "pollutant magnifiers," many organisms concentrate many such toxins, often to their great harm. This is why extreme measures so often are needed to rid our habitats of mere traces of many industrial chemicals.

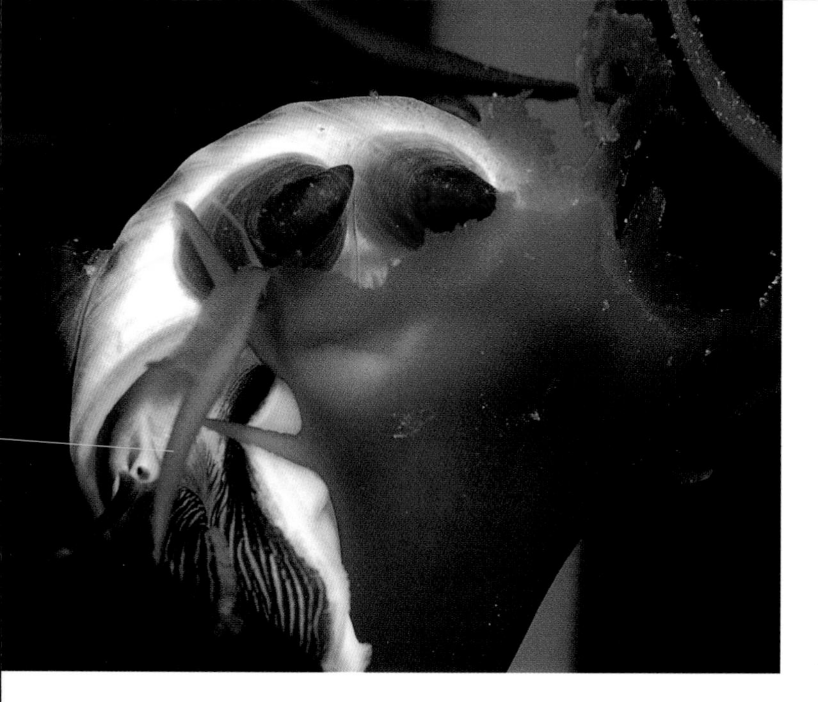

Crepidula adunca
Phylum Mollusca / Class Gastropoda

Young slipper snails, 5 mm, find a host snail. Monterey Bay, 5 m deep.

Two young male slipper snails have climbed onto a large turban snail near the aperture of its shell. The turban snail, apparently unaffected by these riders, is grazing on giant kelp.

Very young slipper snails are mobile. Initially independent grazers, they soon seek a host snail to ride about on, eventually settling on the host's spire and becoming immobile suspension feeders. They use their long gill filaments for both filter-feeding and respiration, and they have no further need to graze.

Slipper snails are sequential hermaphrodites. All begin life as males. At about two months and a size of 10 mm, males start to change into females. The transformation takes several weeks, a time of rapid growth. Then other young males will climb aboard these parvenue females and position themselves to mate. We have found *Crepidula* on nine different species of host snails, each mobile and able to carry its riders to safety when it glides away from seastar predators.

A permanent home site on a topsnail, about 25 mm high. Monterey Bay, 5 m deep.

Now established on its host, this slipper snail is transformed into a suspension feeder and does not budge. Evidence: the rings of the host topsnail, *Calliostoma canaliculatum*, are impressed as precise grooves in the slipper snail's growing shell.

Opposite: A mated pair of *Crepidula* on a topsnail, about 25 mm high. Monterey Bay, 5 m deep.

Two single males and a mated pair (the little male atop the older, larger female) have established their home on the spire of a topsnail, *Calliostoma canaliculatum*. Female *Crepidula* may release a chemical signal that attracts males (Coe 1953). The male has a very long penis with which to reach around and beneath his mate's shell and into her mantle cavity to deposit sperm. Embryonic development and brooding of the young occur in a persistent brood space beneath the mother's foot. *Crepidula* has no free-living larval stage. About 150–200 juveniles, each 2 mm long, are simply pushed out by the mother to begin their independent life (Putnam 1964).

A quick experiment. "Host" about 18 mm.

A convex surface, supplied here by a marble, appears to be one important criterion a slipper snail uses in selecting a new host. We removed several slipper snails from their original hosts and liberated them in an aquarium along with marbles and a small glass cube. Overnight, the females climbed aboard the marble "hosts," ignoring the flat-sided cube, and the males repositioned themselves on the older and larger females.

Doriopsilla albopunctata
Phylum Mollusca / Class Gastropoda

Dorid nudibranchs mating, larger one about 35 mm, and their egg ribbons. Carmel Bay, 24 m deep.

These nudibranchs appear to be much busier mating than feeding. Their egg ribbons lie nearby, along with tube worms, a bryozoan, and crustose coralline red algae. Since nudibranchs are hermaphrodites and mutually copulate, probably both these animals are responsible for depositing the egg ribbons. This species does not have a radula, yet it feeds on several species of sponges. The animal first macerates its food and then sucks out soft tissue, apparently avoiding the sponge spicules, which have not been found in its gut. *Doriopsilla* and a look-alike nudibranch may require dissection for accurate identification.

How on earth do these soft-bodied, shell-less little molluscs defend themselves? Just as nudibranchs such as *Aeolidia* (see chapter 2) acquire nematocysts from their prey for use in their own defense, *Doriopsilla* acquires chemical defenses from its sponge prey. Sponges are notorious for producing toxic compounds, as we have seen, and nudibranchs such as *Doriopsilla* store such compounds in specialized skin glands. Perhaps ancestral shelled molluscs learned to feed on sponges, developing tolerance for their spicules and their chemical deterrents, then learned to store the sponges' chemical deterrents rather then excrete them in feces. Armed chemically, the ancestral snail could shed its shell in the course of further evolution, resulting in today's dorids (Faulkner and Ghiselin 1983).

The nudibranchs' bright yellow color is due to the carotenoid pigment carotene, which occurs in many sponges. In contrast to most nudibranchs, this species' embryos develop within the egg capsules, bypassing the larval stage to emerge as miniature crawling juveniles.

Doriopsilla—and *Hermissenda* too (see next plates)—are used widely in neurophysiological research. Little *Doriopsilla*'s giant neurons, it turns out, have the largest cell bodies of any neuron known: one can identify and isolate single brain cells and study electrical events and the activity of specific chemical neurotransmitters in the course of behavioral experiments. Among the neurotransmitters active in this nudibranch's brain are serotonin, dopamine, acetylcholine, GABA, and nitric oxide—but surprisingly, in light of this long list, not the widespread chemicals epinephrine or norepinephrine (Stuart Thompson, pers. comm.). These days the names of these chemicals appear frequently in the news, in the context of breakthroughs in the understanding of human diseases and new drugs with which to treat them. And here they are at work in a nudibranch snail.

Hermissenda crassicornis (= Phidiana crassicornis)
Phylum Mollusca / Class Gastropoda

Opposite: Eolid nudibranch on compound tunicate, 35 mm. Monterey Bay, 10 m deep.

This beautiful nudibranch is a fairly close relative of *Aeolidia papillosa.* But whereas *Aeolidia* is cryptic, often mimicking its prey in color and in form, *Hermissenda* displays strikingly bold colors—due mostly to carotenoid pigments, passed up the food chain from algae—apparently flaunting its unpleasant taste to potential visual predators. In one study, wrasses, which have excellent color vision, fed readily on *Aeolidia* but could not be induced to take *Hermissenda* (Harris 1987). But since both *Hermissenda* and *Aeolidia* may feed on cnidarians and then store their prey's nematocysts in their cerata, it isn't clear why the wrasses respond differently to the two species. There could be qualitative differences in the two species' stored nematocysts—*Hermissenda,* for example, often preys on strawberry anemones and solitary corals that have particularly potent nematocysts. Or could *Hermissenda,* like other of its eolid cousins, be further protected by noxious secretions?

Hermissenda, a generalist carnivore, is able to chemically detect several hydroid prey organisms at a distance, even without prior contact. It also learns, from previous feeding experience, to recognize and move toward various organisms in other phyla. This one is feeding on a compound tunicate, a frequent prey item. The oral tentacles and rhinophores may be involved in *Hermissenda*'s distance chemoreception (Avila 1998).

Exciting neurophysiological studies on molluscs such as *Hermissenda* may help explain at the most fundamental molecular level what happens when we learn. This nudibranch is a good model to work with because its nervous system, including its brain, has only about five thousand nerve cells. In comparison, the human brain may have a trillion nerve cells, organized with daunting complexity.

Hermissenda can be conditioned to associate two different stimuli, just as Pavlov's dog learned to salivate at the tone of a bell. Instinctively, *Hermissenda* moves toward light, and during rough water it hangs on more tightly. These reactions can be exploited experimentally: a light flashes and simultaneously *Hermissenda* is rapidly rotated, simulating rough water. Soon the light by itself produces the nudibranch's response to rotation: it hangs on more tightly (Matzel 1990). In *Hermissenda,* a single nerve cell lies at the junction between a nerve from the statocyst balance organ (analogous to the human ear's semicircular canals) and a nerve from the eye. What happens when that single cell learns to pair these two events, flash and spin? The neurotransmitter GABA mediates nerve impulses that enter the junction cell from the two sensory organs, eye and statocyst. Within the junction cell, a protein (termed G protein) is modified during conditioning: it "learns." This protein, isolated from one *Hermissenda* shortly after its Pavlovian conditioning and injected into the visual pathway of another *Hermissenda*, transfers memory and produces the learned response to a light flash in an untrained individual (T. Nelson, Collin, and Alkon 1990; Alkon 1993)! Such fundamental observations on "the chemistry of learning" in a simple animal like this may lead to a better understanding not only of learning but also of neurodegenerative diseases in much more complex animals like mammals, including us.

An enigma: are these individuals about to fight or to mate? Open coast, 25 m deep.

Since *Hermissenda* can be cannibalistic, a serious fight may ensue from this encounter as the aggressive little molluscs lunge at each other and bite. Or they could be about to mate. In that more benign scenario, they will touch each other with their tentacles and then slowly glide forward along the right side of the other until their sexual pores are aligned. Then penises evert and mutual copulation follows, lasting for only a second or two. Sometimes one animal misses the other's sexual pore and sperm transfer fails to occur. On such occasions, one nudibranch will turn its head and ingest any semen the other nudibranch left about its own sexual pore (Rutowski 1983).

Strongylocentrotus purpuratus
Phylum Echinodermata / Class Echinoidea

Purple sea urchin, about 5 cm across. Monterey Bay, 17 m deep.

Purple sea urchins often decorate themselves in a manner reminiscent of decorator crabs (see *Loxorhynchus crispatus,* below), opportunistically picking up materials from their environment and attaching—or at least holding—them on their backs. For the urchin, what purpose does decorating serve? For *Strongylocentrotus purpuratus* living in the intertidal zone, decorating may provide some protection from UV light, from desiccation, and from visual predators, though urchins there do fall prey to gulls, black oystercatchers, and crows. In Washington state, in a study to determine the effects of bird predation on purple urchin abundance and, indirectly, on the abundance and diversity of algae, birds were excluded from the intertidal zone by a large cage. The investigator concluded that under normal conditions birds eventually take about half the purple urchins in the intertidal zone, and by taking so many they indirectly increase algal cover and diversity (Wootton 1995).

Subtidally, decorating is less common, and we don't see it there at all among densely massed urchins. This solitary animal has applied a skimpy covering of kelp and a piece of shell, probably not enough to fool a sea otter. But the urchin's protruding spines, its venomous pincers, and its position in the depression it has excavated in the rock do provide a defense. Note the many tube feet that are used in respiratory exchange, locomotion, and probably decorating.

Urchins do not have eyes, not even simple pigment cups, but they do respond to a variety of chemical signals. Some urchins respond to an alarm pheromone; should one urchin be crushed, others nearby will move a meter or two away if they can (Snyder and Snyder 1970). To cite another example, a purple urchin sensing the approach of the predatory sunflower seastar *Pycnopodia* 5–10 cm away responds by waving its spines and pincers and retreating (Dayton 1973).

Since the return of the voracious sea otter to the Monterey Peninsula subtidal sea urchins have become crevice dwellers that avoid exposed sites, obtaining sustenance not from living seaweeds but from detached kelp blades that drift within their grasp. Nevertheless, the urchins in the area are vulnerable, and some sea otter scat consists predominantly of purple urchin fragments as well as mussel remains (Faurot, Ames, and Costa 1986). Additional evidence of otter predation: some otters' bones and the enamel of their teeth are stained purple from the pigment of ingested urchins.

Right: "Urchin-dominated barrens." Open coast, 24 m deep.

In 1986–1987 purple and red urchins (*Strongylocentrotus franciscanus*) virtually denuded a rich reef on the outer coast off the Monterey Peninsula, taking living seaweeds as well as many invertebrates (Watanabe and Harrold 1991). This is a desolate scene from that dramatic event. Purple urchins crowd onto the remnant stipe of the understory kelp *Pterygophora californica;* a sponge and a few *Corynactis* anemones survive, but eventually even these will disappear.

Within two years, however, the urchins themselves were almost all gone, having vanished as mysteriously as they had appeared, and the reef's recovery was under way. The dynamics of this immense ecological event remain especially perplexing because it was so isolated: nearby reefs were unaffected. The mere happenstance of a heavy larval settlement of urchins in 1984 may have initiated the plague.

Sea otters can dive to these depths; why did they not find the massed urchins there? Other predators, disease, loss of forage, and heavy storms may have led to the urchin population's collapse, but these necessarily are vague and speculative hypotheses, as are so many other "explanations" of what is going on in the sea along our shores. After all, similar urchin plagues have occurred in southern California, wiping out vast kelp beds there too—in the absence of the sea otter.

Aboral (top) surface of a young urchin, about 12 mm across. From dock, Monterey Harbor.

In this kaleidoscopic view of an urchin's back, the anus lies at the center, surrounded by spines, pincers, and tube feet. The sieve plate with its perforations that lead to the animal's internal hydraulic system lies here, as do the five genital openings, arranged inconspicuously around the anus. Sea urchins have the basic echinoderm five-part radial symmetry, though it's not as obvious as it is with seastars.

Young urchin, oral (bottom) surface, about 12 mm across. From dock, Monterey Harbor.

This animal's chewing apparatus, an elaborate structure with five pointed teeth worked by dozens of muscles, is called Aristotle's lantern, ever since he likened it to "a horn lantern with the panes of horn left out" (quoted by Milne and Milne 1967). The jaw's five teeth meet in the middle and, tiny though they are, make up for normal wear by growing 1 mm a week. The little white, stalked pincers surrounding the mouth are closed. The elongated tube feet sense the substratum as the urchin creeps along in search of food.

Below: Juvenile urchin on giant kelp, 2 mm spine tip to spine tip. Monterey Bay.

We found this minute urchin in the giant kelp canopy. It had only recently completed its metamorphosis from a larva. Note its many pincers gaping open, each with three jaws (in contrast to seastar pincers, which have only two jaws). This youngster will be lucky to avoid predators and reach the shelter it will need among the rocks below. Many larvae that have the misfortune to settle in the canopy fall prey to señorita fish, *Oxyjulis californica*, which browse invertebrates from kelp blades (Tegner and Dayton 1977; see also chapter 3).

Opposite: The purple urchin as architect: *Strongylocentrotus* skeleton, field of view 30 mm.

This detail of an intact skeleton that measured 9.5 cm shows tubercles, points of attachment for the concave bases of the urchin's spines, both large and small; together, tubercles and spine bases form ball-and-socket joints. Note the tiny paired holes between the tubercles. In the living urchin, each pair of openings contains branches from the water-vascular system and serves one tube foot. The skeleton is made up of mineral plates that are sutured together in lines barely visible in this close-up view. As in our own skull, growth occurs at the suture lines.

Echinoderms are extensively used in embryological experiments, particularly to study fertilization and early development. The egg, a large cell with all the nutrients and half the genetic information needed for the zygote's development, must fuse with a tiny sperm that bears the other half of the genetic material. Amazingly, it wasn't until 1875 that the biologist Hermann Fol, using the seastar's eggs and sperm, observed fertilization under a microscope and for the first time established the role of semen in procreation. It's easy to repeat his experiment: one has only to suspend urchin eggs in seawater and add sperm while stirring, and within seconds the eggs are fertilized. As many as two thousand sperm may bind to one egg's surface, but only one sperm is allowed to enter. Immediately, the egg produces enzymes that destroy all the other sperms' binding sites. All sexual animals have formidable mechanisms to prevent a second sperm's entry. After fertilization, a very, very complex chain of events transforms the egg's surface chemistry, and voilà—development begins (Epel 1976).

Asterina miniata (= Patiria miniata)
Phylum Echinodermata / Class Asteroidea

Bat stars, about 9–12 cm. Monterey Bay, 13 m deep.

Bat stars are omnivores, scavengers, and opportunistic predators. These may be feeding on drift algae, although tube worms, solitary corals, and a variety of other organisms also are potential prey. Additionally, seastars are known to exhibit aggressive behavior toward each other, usually when food gathering (Wobber 1975), and some "arm wrestling" may be going on here. Time-lapse photography of *Asterina* has shown that the animals slowly but surely tear about in a dynamic social interaction that we, living at our much faster tempo, could not otherwise imagine.

One June we saw a dramatic example of bat star reproductive behavior. Several of these seastars were plastered against the window of a large aquarium. Females were emitting flocculent material from pores near the bases of their arms, obviously spawning, and males at a distance of at least 5 m were "smoking," unmistakably releasing sperm. A pheromone coordinates the "follow the leader" responses of the two sexes. A phytoplankton bloom apparently initiated this spawning, as often occurs; at the time, the sea was very cloudy with plankton, assuring food for the larvae—the seastars evidently respond to chemicals coincident with these favorable conditions. And plentiful food is very important during early larval development. *Asterina* larvae reared at high planktonic food levels are larger, more opaque, and better pigmented and calcified; they also develop more quickly than their food-deprived siblings. These changes improve their settlement and metamorphosis success (Basch and Pearse 1996).

Above right: Asterina feeding, about 12 cm across. Carmel Bay, 20 m deep.

This bat star is feeding on the egg ribbon of a dorid nudibranch. Sensory tentacles at the tip of each seastar arm are thought to perceive chemical stimuli and help in the search for food.

Opposite: Asterina captured by the anemone *Urticina lofotensis,* view 18 cm. Carmel Bay, 20 m deep.

Often we have seen *Asterina* being eaten by *Urticina,* as in this photograph. But surprisingly, we have never seen the anemone capture a different species of seastar. At first we thought insensitive scuba divers might have "arranged" the photo op, but we have seen the event occur in the absence of other divers and come to realize it is a natural phenomenon. *Asterina* has fewer tube feet that hang on much less tenaciously than those of many other species

of seastar. In contrast, *Pisaster ochraceus* and *P. giganteus,* for example, cling to their substratum so tightly that tube feet tear off if the seastars are forcibly detached. Thus ocean surge readily dislodges *Asterina* and may sweep it into an anemone's embrace. And could *Asterina*'s lack of defensive pincers make it more attractive to anemones (and to predatory seastars) than seastars that do pinch?

Below: Aboral surface and sieve plate of *Asterina*, field of view about 20 mm.

Asterina often has glorious, variegated colors derived from its ingested carotenoid pigments, as illustrated in this series of photos. Here, the buttonlike, yellow structure is the sieve plate, the perforated opening of the seastar's water vascular system of canals that extends into each arm of the seastar, helping work that arm's tube feet. (The anus, also located aborally, is not evident in this view.) A combination of varying hydraulic pressure and muscular action permits coordinated movements of the tube feet. A seastar's tube feet perform many functions: locomotion, clinging, gas exchange, food capture, and sensory perception.

We have never seen foreign material attached to a seastar's body surface. Beating cilia on *Asterina*'s body surface may create sufficient water currents to clear it of debris, a task performed by the pincers in many species. The seastar's thin, blue skin gills are prominent in this image.

Pycnopodia helianthoides

Phylum Echinodermata / Class Asteroidea

Opposite left: Sunflower seastar, about 40 cm across. Carmel Bay, 20 m deep.

This predator is gliding along among strawberry anemones, encrusting red sponges, compound tunicates, and various bryozoans—altogether a colorful sight. Note that it has lifted several of its arms to clear the anemones, undoubtedly to avoid their stings. This animal, at 40 cm, is big enough, but others of this species may grow to a meter's breadth, the largest of any seastar. With about two dozen arms and thousands of tube feet, these formidable predators are able to overtake and subdue a wide variety of prey, including crabs, chitons, dying squid, turban snails and topsnails, small abalone caught in the open, sea urchins, and other seastars. They may even catch benthic fish. When they are in a hurry, these impressive animals have been clocked at over a meter per minute. But perhaps they tire quickly; observations made both day and night show these seastars at rest in a crack 95 percent of the time (J. Pearse, pers. comm.).

Like *Asterina* and indeed most seastars, *Pycnopodia* reproduces by broadcast spawning of gametes to the sea, a risky business when the animals are dispersed. But it helps that male and female seastars spawn "contagiously," an act coordinated by a pheromone released by one or both sexes. For *Pycnopodia* the chances of fertilization are further enhanced by a sperm chemoattractant in the eggs (Punnett, Miller, and Yoo 1992).

Injured *Pycnopodia* may release an alarm pheromone; uninjured ones, in any case, are sensitive to certain chemical signs of trouble and will flee if fluids and fragments from another *Pycnopodia* are placed nearby (Lawrence 1990).

Although *Pycnopodia* seems to rely principally on tactile information to locate its prey, it has been shown experimentally to detect prey at a distance by chemoreception. For example, in a Y-trough it will move upstream toward both purple and red urchins, not discriminating between the two species at a distance. Neither does it discriminate when thrown into physical contact with the two species: it attacks them both. But the outcome of these attacks differs markedly. Ninety-eight percent of the time *Pycnopodia* successfully captures and ingests purple urchins, unfazed by their tiny, mildly venomous pincers. Its attacks on red urchins, however, almost invariably are foiled; although the red urchin's pincers are not venomous, their long spines prove to be a very effective defense indeed (Moitoza and Phillips 1979).

At the same time, prey animals downstream from *Pycnopodia*, such as purple urchins and turban snails, use their own chemoreception to detect and respond to the big animal's presence. Purple urchins even show a greater defensive response to active *Pycnopodia* than to resting ones. Perhaps the increased hydraulic pressure associated with the seastar's tubefoot motion increases the release of stimulatory chemicals (D. Phillips 1978). When in the presence of a sunflower star, many invertebrates show striking escape responses. For example, topsnails and abalone gyrate violently in an effort to break loose from its clinging tube feet and then rush away; sea cucumbers and the nudibranch *Dendronotus iris* take off swimming; and the purple urchin *Strongylocentrotus* flees.

Above right: Pycnopodia's pincers and skin gills, aboral surface, 17 mm across. Carmel Bay, 20 m deep.

This close-up shows *Pycnopodia*'s two-jawed defensive pincers. They are of two sorts: a few scattered big ones and numerous small ones—several of the latter gaping open—tightly clustered about spines. The pincers probably deter some predators, keep the seastar's surface free of debris, and protect the vulnerable skin gills. Despite the pincers, however, *Pycnopodia* falls prey to another large seastar, *Solaster dawsoni*, that we have found in Carmel Bay at 25 m depth. The thin-walled, densely clustered skin gills and the tube feet on the oral side both aid in gas exchange and in ridding the seastar of metabolic wastes.

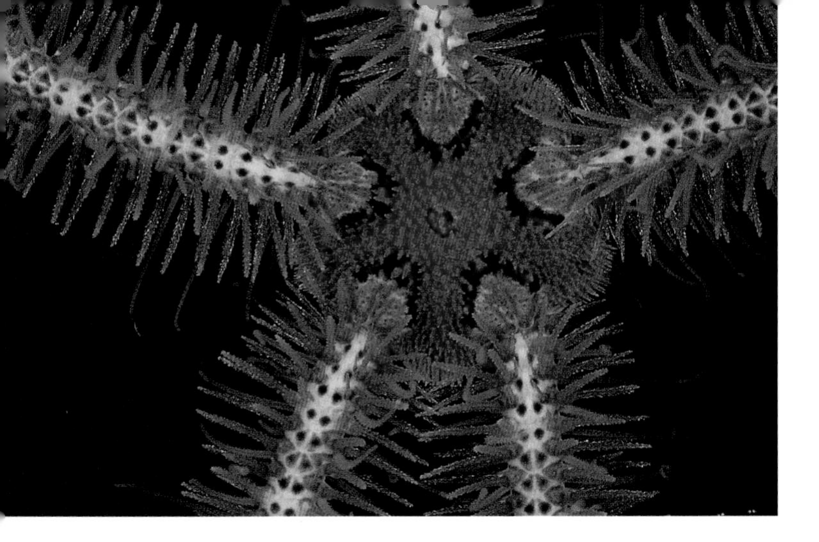

Ophiothrix spiculata
Phylum Echinodermata / Class Ophiuroidea

Brittle star, central disc 10 mm, from giant kelp holdfast. Monterey Bay, 15 m deep.

Brittle stars are agile echinoderms, using their whole arms to creep over the substratum. Most are nocturnal, thereby avoiding diurnal visual predators such as fishes. During the day one is likely to see *Ophiothrix* arms only occasionally protruding from the animals' hiding places in kelp holdfasts, in the sand, and beneath rocks. But once on a 30 m daytime dive on the open coast we saw a thick mat of thousands of these animals out in the open. Could this have been a reproductive congregation? Being in a dense crowd would enhance the chances of fertilization if the brittle stars were synchronously spawning, as many echinoderms do.

Prominent thorny spines and longer, pale tube feet protrude from the sides of the brittle star's arms, as in this photo. The tube feet are pointed and covered with tiny papillae, and they contain sense organs and mucus-secreting glands. Though lacking in suction cups, they help the animal adhere weakly to the substratum as well as aiding in feeding and respiration. The brittle star's mouth and sieve plate are on the underside. Its saccular gut lacks an anus and, together with the gonads, fits inside the little central disc. The animals respire and release waste and gametes through sacs that open through slits at the base of each arm.

During feeding, *Ophiothrix* extends two-thirds of each arm upward, trapping plankton and detritus with its tube feet and with mucus strands on its spines. The animals then march the food particles along to their mouth with the coordinated activity of their very agile tube feet. In another species of *Ophiothrix* with similar feeding habits, tube feet moved a small mass of food along the brittle star's arm at 0.5 mm/sec, and the mass grew in size, trapping more particles before it reached the mouth (Warner and Woodley 1975).

Brittle stars are so called because their arms readily break or detach if seized. The animals regenerate these missing parts, while the predator is left with a writhing limb that is mostly skeletal blocks and spines of calcium carbonate.

Opposite: Different color patterns.

Brittle stars show beautiful color patterns of metabolically altered carotenoid pigments derived from ingested phytoplankton. The variegated patterns seem elaborately fragmented, enough, perhaps, to confuse visual predators.

Loxorhynchus crispatus
Phylum Arthropoda / Class Crustacea

Right: Spider crab, shell 9 cm across. Carmel Bay, 15 m deep.

Some spider crabs are called "decorator crabs" because of their habit of affixing various materials to their shells. This relatively large decorator attached a bit of white colonial tunicate—which has grown robustly, partially covering the crab's back—and it has skimpily planted its legs with a few sprigs of red algae. Smaller spider crabs decorate much more heavily and are rarely as conspicuous as this one, which stands out against a colorful background of bright *Corynactis* anemones and lacy and fluted bryozoans. *Loxorhynchus* is a generalist that feeds on drift kelp and both living and dead invertebrates.

To apply its decor, the crab picks up morsels with its claws, roughens their edges with its mouthparts, and reaches deliberately about its body to wedge its preparations among the tiny hooked bristles that cover its back and legs. Other sensory bristles inform the crab about the status and position of its decorations. Following a molt, the crab must repeat the process, often using decorations gathered from its cast-off exoskeleton. *Loxorhynchus* crabs readily apply adornments that do not match their immediate surroundings (they even decorate with their eyes covered; Wicksten 1993), but these mistakes scarcely belittle the underlying ingenuity of their decorative obsession.

Why decorate? It scarcely can fool their prey, which are mostly algae or sedentary animals. It could, however, help defend against visual or tactile predators such as fish, octopuses, and sea otters. Even if its decorations do not make *Loxorhynchus* invisible, they certainly make it look less crablike; and its habit of remaining immobile during the day doesn't hurt either. During dives, we repeatedly have failed to recognize a decorator crab until it sees us and abruptly moves. Characteristically, during their flight these crabs scamper off a ledge and parachute down with legs outstretched. To an octopus, *Loxorhynchus* probably does not even feel like a crab, and its decorations often are ones that few marine animals eat, such as noxious sponges and hydroids.

But does *Loxorhynchus* really need to decorate? The evidence is inconclusive. Large *Loxorhynchus*—some measuring almost a meter across the outstretched legs—stop decorating; they probably have few predators. In an aquarium, wolf-eels spat out decorated young *Loxorhynchus* or ignored them. Curiously, observations in the field and in aquaria suggest that in general sea otters and fishes such as rockfishes, sculpins, cabezons, and California sheephead ignore or reject *Loxorhynchus* crabs whether or not they are decorated. In several aquarium trials between a red octopus and both decorated and stripped *Loxorhynchus*, only one crab was eaten, and it was a decorated one. Nevertheless, the flesh of spider crabs is edible—fish and bat stars promptly strip the flesh from dead ones. These various predators also readily consume crabs in other families, such as kelp crabs and cancer crabs. We are left with the uncomfortable notion that while decorating should confer all kinds of benefits, *Loxorhynchus* seems not to be a very vulnerable species, with or without its decorations. Wicksten (1993) presents the evidence and gives an excellent discussion of these complex issues.

Loxorhynchus crab, shell 5 cm across. Carmel Bay, 18 m deep.

We often fail to notice a heavily decorated crab like this young one until a piece of the seafloor gets up and walks away. This animal is facing toward the right. Once you grasp the image, you can see the crab has attached articulated coralline algae, erect and branching bryozoans, and many *Corynactis* anemones. Many of these same organisms are in the crab's immediate background. It's a costume fit for a champion!

Podochela hemphilli

Phylum Arthropoda / Class Crustacea

Hemphill's kelp crab, shell 25 mm across. Monterey Bay, 14 m deep.

Podochela is still another crab that decorates itself. Often it attaches bits of red algae, concentrated on the outer half of its first pair of walking legs. This crab has added hydroids to the array. While this little animal is easy to see perched on the yellow bryozoan *Phidolopora pacifica*, sometimes these crabs seem to disappear completely into a background of seaweeds. They compound the camouflage by making bowing motions that wave their attached algae back and forth, simulating the sway of seaweeds in the ocean's surge (MacGinitie and MacGinitie 1968).

How did the decorating habit evolve? Mary Wicksten (1993) points out that the first motions of decorating resemble those of feeding: the crab plucks food with its pincers and carries it to its mouthparts. It may not have been a big evolutionary step for crabs of this sort to store uneaten food by attaching it to the hooked bristles on their backs, the serendipitous result being a sort of camouflage. When *Podochela* is deprived of its regular food sources, it readily eats its own attached algae, as though it had stored them for just such an emergency.

Artedius corallinus
Phylum Chordata / Subphylum Vertebrata / Class Osteichthyes

Coralline sculpin, 9 cm long. Open coast, 25 m deep.

These sculpins are so well camouflaged that they are almost impossible to see until they suddenly move a little. This one is obvious as it lies between a sponge and some partially closed *Corynactis*. Notice that its head is as yellow as the sponge beneath it. This is an excellent example of camouflage by means of chromatophores, pigment-containing cells in the skin. The cells' pigment granules may be either red, orange, yellow, or black; different shades are produced by blending. Control of the chromatophores apparently is mediated by neurotransmitting chemicals released by neurons that innervate these cells, allowing the fish to alter its color and pattern quickly. The cells' different pigment granules can either be aggregated to form inconspicuous spots, or they can be dispersed to give overall color to the chromatophores. This sculpin, stimulated by what it has seen of its surroundings, may have dispersed the yellow pigment granules within chromatophores in its head while aggregating those of other colors. Should it now move to a dark substrate and look about again, the yellow pigment granules will aggregate and others will disperse; the head will then darken and look like the rest of the body. These rapid color changes, involving the aggregation or dispersal of pigments, are "physiological," as opposed to "morphological" color changes brought about by a slow color change in the pigment granules themselves.

The large head and fanlike pectoral fins of this fish are sculpin characteristics. These bottom-dwelling fishes lack swim bladders; unable to adjust their buoyancy, they sink when they stop swimming. Sculpins copulate, and females lay eggs in a nest that the male guards.

Sebastes miniatus
Phylum Chordata / Subphylum Vertebrata / Class Osteichthyes

Vermilion rockfish, about 40 cm long. Carmel Bay, 18 m deep.

The vermilion is one of dozens of rockfish species that inhabit California's coastal waters. The larger vermilions live offshore at depths below 60 m; this young fish, identifiable as such by its red fins tinged with black, was hovering under a ledge in relatively shallow water. It would appear gray rather than vermilion or red were it not illuminated by a strobe. Vermilions are an important part of the commercial rockfish catch, often taken by offshore trawls in deep water. In California, trawls are prohibited within three miles of shore (to reduce sea otter mortality), and minimum size mesh is limited to 4½ inches, providing an escape for juvenile fishes. Rockfish often are served in local restaurants as red snapper, even though snappers are in a different family. Rockfish can be aged by examining the rings in their scales, or by their ear bones. Vermilions can attain up to 75 cm length and live as long as twenty-two years. The adults generally feed on octopuses, squid, and small fishes such as anchovies and blue lanternfish, though at times they take larger planktonic organisms as well (J. Phillips 1964).

Young-of-the-year *Sebastes miniatus,* about 6 cm long. Carmel Bay, 18 m deep.

Vermilions bear their young as larvae during the winter. The larvae, which start out translucent, spend several months offshore in the plankton feeding, acquiring pigment, and developing before they descend as young-of-the-year to a bottom-dwelling existence. Note how strikingly different this juvenile's appearance is from that of the adult. The blotchy coloration probably provides camouflage during the fish's sojourn on a sandy, rocky, algal-strewn seafloor. Note, too, the tinge of vermilion on the fins' fringes, an early change toward adult coloration. It's characteristic for these juveniles to rest at the head-up angle depicted here.

Artedius corallinus
Phylum Chordata / Subphylum Vertebrata / Class Osteichthyes

Coralline sculpin, 9 cm long. Open coast, 25 m deep.

These sculpins are so well camouflaged that they are almost impossible to see until they suddenly move a little. This one is obvious as it lies between a sponge and some partially closed *Corynactis*. Notice that its head is as yellow as the sponge beneath it. This is an excellent example of camouflage by means of chromatophores, pigment-containing cells in the skin. The cells' pigment granules may be either red, orange, yellow, or black; different shades are produced by blending. Control of the chromatophores apparently is mediated by neurotransmitting chemicals released by neurons that innervate these cells, allowing the fish to alter its color and pattern quickly. The cells' different pigment granules can either be aggregated to form inconspicuous spots, or they can be dispersed to give overall color to the chromatophores. This sculpin, stimulated by what it has seen of its surroundings, may have dispersed the yellow pigment granules within chromatophores in its head while aggregating those of other colors. Should it now move to a dark substrate and look about again, the yellow pigment granules will aggregate and others will disperse; the head will then darken and look like the rest of the body. These rapid color changes, involving the aggregation or dispersal of pigments, are "physiological," as opposed to "morphological" color changes brought about by a slow color change in the pigment granules themselves.

The large head and fanlike pectoral fins of this fish are sculpin characteristics. These bottom-dwelling fishes lack swim bladders; unable to adjust their buoyancy, they sink when they stop swimming. Sculpins copulate, and females lay eggs in a nest that the male guards.

A well-camouflaged *Artedius corallinus*, 8 cm long. Monterey Bay, 13 m deep.

Even caught and framed in a photograph, this little sculpin is difficult to discern against the confusing, busy background. For the moment, it has curled around the shell of a turban snail that a hermit crab occupies.

Gibbonsia, probably *elegans*

Phylum Chordata / Subphylum Vertebrata / Class Osteichthyes

Below: Spotted kelpfish (?), 7 cm long. Monterey Bay, 14 m deep.

There are several look-alike *Gibbonsia* species; it takes a microscope to tell them apart for sure. Like sculpins, they are bottom-dwelling fish that lack swim bladders and so cannot adjust their midwater buoyancy. They swim briefly and then sink to the bottom and "disappear" by blending in with the algae around them.

Gibbonsia occurs in three color morphs: red, green, and brown. This one closely matches the surrounding coralline red algae. To conform with seasonal changes in the algal background, these fish gradually alter both the color of the carotenoid pigment granules in their chromatophores and the number of chromatophores. In the field, it took *Gibbonsia* about ten days to accomplish these "morphological" color changes. Kelpfish cannot synthesize carotenoids *de novo* but acquire them from their prey, predominantly alga-grazing amphipod and isopod crustaceans. Most of these crustaceans match the color of their algal forage.

Studies of kelpfish gut contents, however, show that the majority of the crustaceans they catch do *not* match the algae where the fish themselves are living and feeding. In other words, prey that deviate in color from their algal background are more vulnerable to capture. So *Gibbonsia*'s three color morphs probably are determined by their perception of algal background color alone and not directly by their crustacean diet (Stepien, Glattke, and Fink 1988).

In addition to the gradual morphological color changes they accomplish by modifying carotenoid pigments, kelpfish are able to "physiologically" alter chromatophores containing black melanin granules through aggregation or dispersal (like the sculpin *Artedius*). By this means they can lighten or darken their bodies within minutes.

Like its prey, *Gibbonsia* benefits from being camouflaged. In behavioral experiments, larger predatory fish were more likely to capture *Gibbonsia* when it did not match its background color.

Scorpaenichthys marmoratus

Phylum Chordata / Subphylum Vertebrata / Class Osteichthyes

Below: Male cabezon, about 45 cm long. Carmel Bay, 20 m deep.

Cabezons, or marbled sculpins, are the largest of our local sculpins. They lack scales, and they have skin flaps over their snout and eyes. The great majority of males are red, while females generally are green. This male must be very confident in his camouflage to allow his portrait to be taken. Adults often feed on shelled organisms such as cancer crabs, abalone and other snails, and chitons, but they reject decorator crabs, *Loxorhynchus crispatus* (Wicksten 1993). Palatable crabs they swallow whole, then spit out the exoskeletons; when the shells are regurgitated, they are highly polished by stomach acids. Cabezons also take squid, octopus, and fish.

Cabezon eggs on the hydrocoral *Stylaster californicus*, field of view about 20 cm. Carmel Bay, 23 m deep.

Cabezons spawn in the fall and winter; we found this egg mass during a November dive. The female usually deposits her eggs on a rocky bottom, where a male then fertilizes them, but she may choose a kelp holdfast or even, as here, the branches of hydrocoral as a nest site. Some 50,000–100,000 eggs may be concentrated in a single nest. Although the fish itself is tasty, the roe is highly toxic to mammals and birds. In the 1920s, an ichthyologist (of all people!) and his wife became quite ill after dining on cabezon roe (Love 1991). So the eggs' visual conspicuousness is made up for by their chemical defenses, and eggs in the open like these survive to hatching even in shallow water. Males usually guard the nest, which would seem to suggest a certain amount of predation on this enticing resource.

Cabezon larvae are pelagic. Surface trawls within 30 miles of the Oregon coast immediately before the upwelling period turned up large numbers of larval and juvenile cabezon, together with other fish species. These fishes occurred in discrete groups associated with zooplankton, particularly dungeness crab larvae, and they were found almost exclusively in the night trawls (Shenker 1988). Their zooplankton prey may be vertically migrating species. Young-of-the-year juvenile cabezon move inshore to tidepools when they are about 5 cm long. Later, as adults, cabezon appear in deeper water.

Lophogorgia chilensis
Phylum Cnidaria / Class Anthozoa

Colonies of red gorgonian coral, about 35 cm high. San Clemente Island, 13 m deep.

The gorgonian, or horny coral, *Lophogorgia* lives in exposed habitats of Carmel Bay, starting at about 33 m deep and becoming prolific below 57 m. Farther south, where this photo was taken, gorgonians occur in shallower water. Since the generally accepted depth limit for recreational diving is 40 m, it takes specialized training and equipment to penetrate safely to depths where many of these animals live.

Gorgonians are octocorals—characterized by eight feathery tentacles on each polyp—with a moderately flexible internal skeleton of horny material and some further strengthening by calcium carbonate spicules. The retractable polyps are borne on branches that may lie in one plane, as in most sea fans, or, as in *Lophogorgia*, may spread out and become bushy. Their skeleton is covered with sheets of red fleshy skin and mucus. The colony constantly sheds its skin and spicules, and these are found in the feces of many grazers even though gorgonians are generally unpalatable, particularly to fishes. Gorgonian flesh is rich in steroids, antibiotic compounds, and the neuromuscular toxin known as lophotoxin. The toxin irreversibly blocks the transmission of impulses from nerves to muscles, even though the muscles themselves can still respond to direct electrical stimulation (Fenical et al. 1981). Cobra venom has a similar action. Toxins that block motor activity have clear advantages both for predation and for defense, in that they immobilize the target.

Just as the aggregating anemone, *Anthopleura elegantissima,* recognizes nonclonemates with the touch of a tentacle, gorgonians recognize self and non-self. A piece of one *Lophogorgia* grafted onto a different species results in non-self reactions that mutually destroy tissue within four days. A graft from one colony to another of the same species shows a

Lophogorgia polyps, branch 12 mm long. Carmel Bay, 34 m deep.

similar rejection but a delayed one—still a non-self reaction. Grafts from one part of a colony to another part of the same colony almost always fuse; they are recognized as "self" (Theodor 1970). These reactions of cellular immunity, such important problems today in organ transplantation, appear to be fundamental properties that appear in simple animals from many phyla.

In contrast to flourishing colonies of *Lophogorgia* in Carmel Bay, hundreds of colonies that appeared healthy in the more sheltered habitats of Monterey Bay at 46 m in 1993 were dead or dying of unknown causes two years later. On the same reefs where this mysterious disaster occurred, gorgonians reappeared in 1996 (Phil Sammet, pers. comm.).

Surge can extend even to this depth, sweeping the gorgonian's tentacles from side to side. The polyps often retract in quiet water and reextend in currents. Here we see the typical octocoral pattern of eight feathery tentacles.

Living on gorgonian colonies both alive and dead is a diverse and little-known community of animals that includes attached suspension feeders as well as roving predators. The following pages depict a few of these mutual inhabitants.

Neosimnia barbarensis

Phylum Mollusca / Class Gastropoda

Opposite: Ovulid snail, 20 mm long, on *Lophogorgia chilensis.* From Carmel Bay, 58 m deep. (Specimen courtesy of Mike Guardino)

Neosimnia snails are specialists that prey only on octocorals (Theodor 1967; Patton 1972). This one is gliding along, its proboscis extended, feeding on gorgonian flesh despite the chemical defenses that *Lophogorgia* and many other octocorals present. Although fish avoid eating most gorgonians, they readily take one species of *Neosimnia—N. uniplicata—* which appears not to acquire its gorgonian prey's antipredator defenses (D. Gerhart, Rittschof, and Mayo 1988). We do not know whether or not our local species is chemically defended. The snail's beady little eyes, lying at the base of the tentacles, consist of a cornea, lens, and sensory cup. *Neosimnia* keeps its mantle raised most of the time, as in the photo. It provides particularly good visual camouflage when the coral's polyps are extended. By grazing on gorgonian flesh, the snail exposes the colony's horny internal skeleton, and this invites other animals to attach themselves as larvae to that nontoxic substratum.

Neosimnia has internal fertilization. Females attach clusters of white eggs in clear, round capsules to the gorgonian flesh. After going through four developmental stages within the capsule, feeding larvae with pinkish-brown shells emerge and swim away (Main 1982).

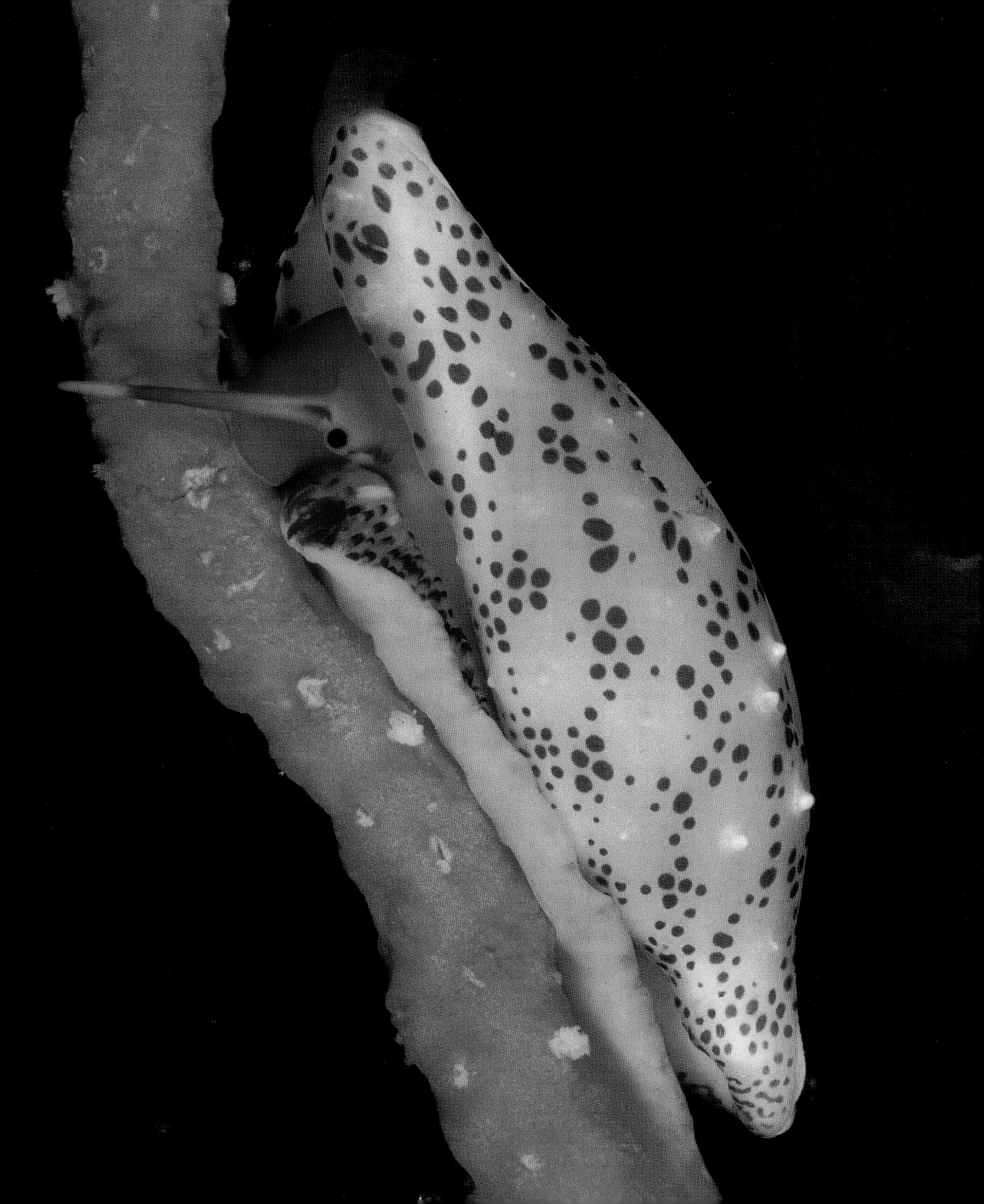

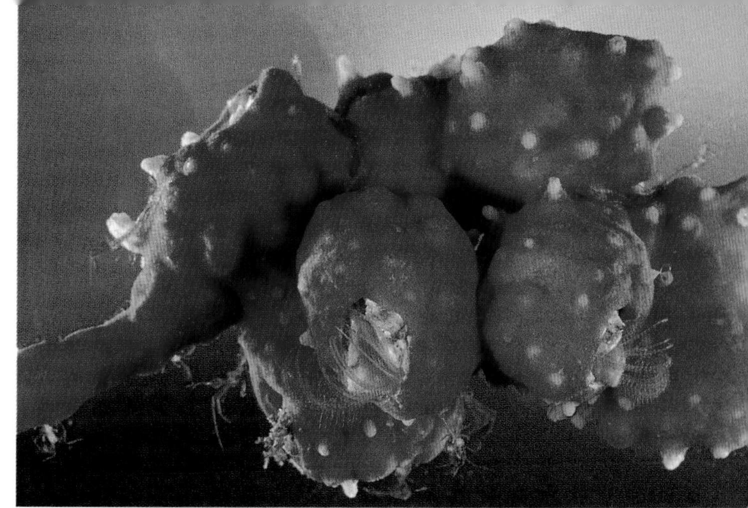

Zoanthid anemone
Phylum Cnidaria / Class Anthozoa

Undescribed zoanthid anemone, about 15 mm long, on dead gorgonian, *Lophogorgia chilensis.* From Monterey Bay, 47 m deep. (Specimen courtesy of Phil Sammet)

Zoanthids, like the small anemones they resemble, do not have a hard skeleton, and their body cavity is partitioned by curtains of septa. Here we can peek inside one of these animals through its translucent oral disc. This zoanthid is an undescribed species, currently awaiting study and as of 1999 unnamed.

Although some zoanthids are solitary, most are colonial, as this species appears to be. Note the cluster of four interconnected polyps. These polyps cling to *Lophogorgia*'s horny, exposed, axial skeleton both on living colonies in Carmel Bay and on dead ones in Monterey Bay. Does an attractant in gorgonian skeletons beckon their larvae? A zoanthid species in southern California, *Parazoanthus lucificum,* also lives on gorgonians. Such specific substratum choice is common among the colonial zoanthids: other species habitually grow on a sponge, a hydroid, a coral, a bryozoan, worm tubes, even shells inhabited by hermit crabs (Hyman 1940).

Balanus galeatus (= Conopea galeata)
Phylum Arthropoda / Class Crustacea

Commensal acorn barnacles on *Lophogorgia chilensis,* each barnacle 10 mm across. From Carmel Bay, 58 m deep. (Specimen courtesy of Mike Guardino)

The adults of this remarkable, very specialized barnacle live only on gorgonians. In our photo, some eight of them are clustered as "galls" or red lumps, attached to *Lophogorgia*'s axial skeleton and overgrown by its soft outer tissue. Several of them are extending their thoracic appendages, fishing for plankton. Note, too, the tiny heads and antennae of partially hidden skeleton shrimps.

In the laboratory, if a living gorgonian is present, the barnacle cyprid larvae avoid its flesh, settling in clusters only where the axial skeleton has been exposed. The cyprids appear to "pace off" a few millimeters from the flesh before attaching. If too close, they may get smothered by the regenerative gorgonian flesh before they become established; if too far away, they may be eaten or fouled before a protective coat of new gorgonian flesh covers them (Lewis 1978).

How can they "anticipate" these potential problems? Simple: This is an example of behavior mediated by substances in the gorgonian flesh that inhibit larval settlement and substances in the exposed axial skeleton that induce it (Standing, Hooper, and Costlow 1984). Once some cyprids have settled and metamorposed, others join the party, settling on the plates of established barnacles. The stimulus for these cyprids to settle gregariously is the presence of "essence of barnacle," arthropodin, a chemical signal from previously settled individuals of the same species (Crisp 1990), not from the gorgonian itself.

The occurrence of many acorn barnacles of another species, *Balanus nubilis*, on the axial skeleton of dead *Lophogorgia* in Monterey Bay may be opportunistic rather than due to an inducer. This species populates widely different sites in addition to *Lophogorgia*.

One quarter of *B. galeatus* larvae are genetically programmed to grow into short-lived, nonfeeding, dwarf males that attach to the plates of the settled hermaphrodite barnacles. These males soon die but are continually replaced, ensuring male partners for any isolated hermaphrodites (Molenock and Gomez 1972; Lewis 1978). Darwin (1854) was fascinated by the phenomenon of dwarf male barnacles: "Some of the males are rudimentary to a degree, which I believe can hardly be equalled in the whole animal kingdom; they may, in fact, be said to exist as mere bags of spermatozoa. So widely do some of them depart in every character from their class, that twice it has happened to me to examine specimens with a little care, and not even to suspect, until a long period afterwards, that these males were Cirripedes."

Celleporina robertsonae *(= Costazia robertsonae)*

Phylum Bryozoa / Class Gymnolaemata

Below: Bryozoan on skeleton of *Lophogorgia chilensis*, the barnacle *Balanus nubilis,* the purple-ringed topsnail *Calliostoma annulatum*, 20 mm high, and the polychaete worm *Serpula vermicularis.* From Monterey Bay, 47 m deep. (Specimen courtesy of Mike Guardino)

The bryozoan *Celleporina* often grows on rock or on algae. It finds a special niche, however, by heavily colonizing the denuded axial skeleton of *Lophogorgia* in both Carmel and Monterey Bays. Is *Celleporina* settling there opportunistically, or could its larvae be attracted by the same "inducer" that beckons *Balanus galeatus* larvae?

These stubby, fingerlike, heavily calcified bryozoan

colonies fall prey to the beautiful purple-ringed topsnail, as seen here, and also to the blue topsnail, *Calliostoma ligatum*. Both species of snail were swarming over our specimens. Under attack, the bryozoan has withdrawn its tiny tentacular crowns. The busily feeding acorn barnacle, *Balanus nubilis,* is still another member of this very crowded community, as is a tube worm whose operculum occludes its tube. Competition for space is so severe—or the gorgonian's "inducer" is so effective—that scarcely a millimeter of gorgonian skeleton remains exposed.

Heptacarpus flexus

Phylum Arthropoda / Class Crustacea

Female "broken back" shrimp, 30 mm long, on *Lophogorgia chilensis.* From Carmel Bay, 58 m deep. (Specimen courtesy of Mike Guardino; identification by G. Jensen)

This mother shrimp is brooding eggs that are attached to her abdominal appendages, appendages which she uses for swimming as well as for brooding. After females molt and become receptive, males deposit sperm packets near their ovipores and fertilization occurs internally shortly thereafter. Shrimp larvae molt repeatedly and metamorphose twice during their sojourn in the plankton.

Although this shrimp is resting at a very exposed site, her translucent body and scattered spots of red and yellow pigment make her relatively inconspicuous. She may be scavenging on spicules, mucus, and

even toxic tissues from the surface of the gorgonian, judging by the red material in her gut. Such materials, largely avoided by fishes, may by their very presence give her an additional measure of protection. Probably due to her feeding, many of the gorgonian's polyps have withdrawn.

The next time you eat shrimp salad, recall to mind the delicate beauty of the living animal now dead on your plate.

Above: Close up, the shrimp's compound eyes. Each 1 mm across.

In contrast to most compound eyes, including those of *Ampithoe* sp. (see chapter 1), this shrimp—along with several other crustaceans—has visual cones, or ommatidia, with square facets rather than hexagonal ones, as clearly shown in the photo. In addition, light is focused on the retina by internal reflections from the sides of *pyramidal* ommatidia, rather than by passing down through a crystalline cone as in the compound eyes of most crustaceans. With her stalked compound eyes and their huge number of ommatidia, *Heptacarpus* should successfully recognize a predator's approach from almost any direction.

As for prey, she probably does not seek it visually. Rather, *Heptacarpus* undoubtedly utilizes olfaction mediated by tiny sensory "hairs" on the first antennae, the little horns that stick up between her stalked eyes. One species of *Heptacarpus* shrimp, *H. pictus*, has been observed to spend much time preening these antennae (as well as other body parts) with its third thoracic appendages, apparently to prevent fouling by organisms such as diatoms and detritus, which would impair this important sensory modality (Bauer 1977).

The shrimp's second antennae, the long ones that arise just beneath the eye stalks and sweep to the rear, serve the sensation of touch.

Metacaprella anomala
Phylum Arthropoda / Class Crustacea

Female caprellid amphipod, 16 mm long, with young, on *Lophogorgia chilensis.* Carmel Bay, 58 m deep. (Specimen courtesy of Mike Guardino; identification by P. Slattery)

This female skeleton shrimp is lying flat on a gorgonian branch, covered and surrounded by a swarm of over two dozen young, almost certainly her brood. Her brood pouch, the light red sac beneath her body and pressed against the gorgonian, appears to be empty. Caprellids, as we saw in chapter 2, are bizarre amphipods with clinging legs at both ends of their body, a greatly elongated thorax, and a reduced abdomen. They climb about like inchworms and swim with undulating body movements.

In some species of caprellids, the juveniles disperse as soon as they are liberated from their mother's brood pouch. Others receive varying degrees of maternal care that may result in faster growth and a higher rate of survival (Aoki 1997). There can be no doubt from these juveniles' behavior that their mother plays a crucial role in their early lives. They may detect her presence with their chemosensory antennae and thus not stray too far. As additional protection, these little transparent juveniles are simply very difficult to see. The mother's deep maroon color (black in the dim light where she was found) is good color camouflage against visual predators like fishes. She may have acquired the pigment scavenging the gorgonian's sloughed-off debris; she could even be chemically protected by the gorgonian's toxins.

Parapagurodes hartae
Phylum Arthropoda / Class Crustacea

Hermit crab, about 5 mm long, on *Lophogorgia chilensis.* From Carmel Bay, 58 m deep. (Specimen courtesy of Mike Guardino)

Several of these very small hermits were climbing over our gorgonian specimens. For a sense of scale, consider that this gorgonian branch is only 2 mm in diameter.

This species of hermit crab was first described in 1996 (McLaughlin and Jensen 1996). It prefers cold water. Although it may occur in shallow water in British Columbia, in southern California it is found only at depths of several hundred meters. The vivid coloration is characteristic of the species. Note the stalked compound (ommatidial) eyes with green and black corneas, which provide an excellent warning system. Note the violet patches on the claw-bearing legs, a characteristic of the species. The smaller, left claw of hermits is used to convey food to the mouth; the larger, right one holds large pieces of food, is used for defense, and blocks the shell opening like an operculum when the crab withdraws for refuge.

The hermit on the left is a male, as revealed by his markedly elongated, claw-bearing right leg. Should the other crab be a receptive female, the male will grasp the shell aperture of his potential mate with his left claw and carry her about, snail house and all, for days until she molts, while warding off other males with his right claw (Jensen 1995).

These hermits have found home in the shells of the snail *Nassarius mendicus cooperi*, which often occur in deep water.

6 The Sandy Seafloor

Most of Monterey's seafloor is covered by soft sediments, much of it sand. At first glance life in these habitats may seem sparse, but actually many organisms are adapted to these shifting environments. Some animals find shelter by living in tubes or burrows; others attach themselves to buried rocks. Species of crabs and snails live largely beneath the sand, emerging only at night. Clams lie permanently buried with their siphons extended to the sediment's surface. Sea pens anchor themselves in the sand with an expanded bulb. Flatfish await their prey by lurking half-buried. These are the large inhabitants; smaller ones work their way through the sand, and microscopic ones live in the water between the sand grains. It is a fully exploited substratum.

Ocean surge may lift clouds of silt and sweep drift algae about, reducing visibility but providing food for suspension-feeding animals. Although unstable sand is unsuitable for most holdfast-dependent algae, beds of rooted eelgrass do crop up in shallow areas of calm water and provide additional habitat for small animals.

Moonglow anemone *Anthopleura artemisia* and the sand-dwelling community.

Polyorchis penicillatus
Phylum Cnidaria / Class Hydrozoa

Hydromedusa, bell about 30 mm tall. Monterey Bay Aquarium.

During the late 1980s large numbers of these beautiful, pulsating little hydromedusae were found in the eelgrass beds near Monterey's Wharf II. Often they were feeding on dense clusters of tiny crustaceans called mysids. What a thrilling sight! (This one was photographed in an aquarium, not quite as exciting.) *Polyorchis* also have been found near eelgrass in Bodega Bay and in Puget Sound. Sadly, this and other species of *Polyorchis*, usually inhabitants of protected bays and inlets, have grown rarer with the years, such that they should be considered threatened now. Their decline may be due to degradation and loss of some of their crucial habitats (Mills 1997).

Note the short, thick filaments hanging from the nutritive radial canals near the top of the bell; these are the gonads. The mouth stalk, or manubrium, dangles to just below the bell's edge. Red eye-spots lie at the base of the tentacles; they have light-sensitive retinal cells but no lens. The animal's misty blue cast is not a pigment but rather a structural color, created by short wavelengths of light reflected from colloidal particles in colorless tissue. *Polyorchis* repeatedly swims upward and then rests and drifts downward, trailing its tentacles around its bell, fishing as it sinks. This one has just started its descent. It seizes its prey with a tentacle, reels it in toward the bell's margin, and then swings its manubrium over to retrieve the meal. Alternatively, *Polyorchis* rests on the bottom, perched on its tentacles, while it feeds on bottom-dwelling organisms by extending its manubrium to the sediment. Periodically it hops off the bottom with a single pulsation, stirring up the sediment, and then settles down again in a new place—altogether unusual behaviors for a medusa (Mills 1981).

Polyorchis does not have a brain, but rather the simple nerve net typical of cnidarians. Here and there some concentrations of nerve cells act as pacemakers for the medusa's repetitive swimming contractions. The neurotransmitter dopamine, occurring in the nerve-rich tissues in the margin of *Polyorchis*, inhibits the hydromedusa's swimming (Chung 1995). This is another instance in which a molecule that is active in an invertebrate's nervous system also plays a role in our own nervous system. In humans, Parkinson's disease, with its uncontrolled tremors, may be due at least in part to a deficiency of dopamine in a layer of cells (substantia nigra) in the midbrain. Additionally, in rats and baboons exposure to addictive drugs like nicotine has been shown to be associated with a surge of dopamine in the brain's "reward centers" (Wickelgren 1998). It may prove possible to block this dopamine surge pharmacologically, thus helping smokers to control their intense nicotine craving.

Presumably, *Polyorchis*, like many other hydrozoan medusae, has an alternate sessile hydroid (polyp) stage in its life cycle. However, where the polyps grow in nature remains a profound mystery (a purported finding some years ago ultimately proved erroneous). Searches of eelgrass beds and the surrounding sand, where *Polyorchis* so often is found swimming, have yielded no clue of a benthic stage. By breeding *Polyorchis* hydromedusae in the laboratory, investigators have obtained their planula larvae, but even their best efforts have not persuaded the planulae to cooperate and metamorphose into polyps. That challenge remains.

Olivella biplicata

Phylum Mollusca / Class Gastropoda

Right: Purple olive snails mating, 25 mm long. Monterey Bay, 5 m deep.

During the day these snails lie hidden just beneath the surface of the sand in shallow water, often in immense numbers, usually emerging from the sand and actively crawling about only at night. This shiny, mating pair, however, surfaced unexpectedly for its picture during the day. In order to copulate, a male will follow a receptive female's mucus trail and, upon catching up with her, attach his foot to her shell with sticky mucus. After mating, she soon deposits quantities of egg capsules on rocks or shells. The embryos develop to the larval stage within the capsules and then hatch and swim away.

The many empty *Olivella* shells that litter beaches often bear clues to how their owners died. If a hole near the shell's apex has straight, parallel sides, the snail probably was eaten by an octopus. If the hole is countersunk with beveled sides, the predator was the moon snail, *Polinices lewisii.* The cone snail *Conus californicus* simply swallows small purple olives whole.

The short-spined, rose-colored seastar *Pisaster brevispinus*, another neighbor of *Olivella*, is one for the snail to avoid as well. If the seastar's tube feet touch them, purple olives exhibit a violent avoidance reaction, rearing and swimming and then gliding away to bury themselves in the sand. In contrast, these snails do not react to the mussel's great enemy, the ochre seastar (*Pisaster ochraceus*) of rocky shores (Edwards 1969). Evidently the snail's chemical sense can distinguish between these closely related seastar species.

Tectura depicta
Phylum Mollusca / Class Gastropoda

Painted limpet, 2 mm long, on the eelgrass *Zostera marina*. Monterey Bay, 10 m deep. (Specimen courtesy of Dr. Richard Zimmerman)

This little limpet appears to graze solely on eelgrass, *Zostera marina. Zostera* is a flowering plant that grows on shallow muddy and sandy bottoms. It's in the same family as the surfgrass *Phyllospadix* of more wave-swept habitats. These seagrasses form the basis for some of the most productive ecosystems on earth.

In the recent past the limpet was known only from southern Baja California north to Santa Barbara County, though it had been found in Monterey County in the nineteenth century, possibly a period of warming. In late 1993, however, *Tectura* reappeared in Monterey Bay, possibly reintroduced by shipping. Since then it has devastated a previously large, unbroken meadow of eelgrass off Del Monte Beach, Monterey, wiping out most of the plants and reducing the meadow to a series of small patches. As many as a dozen limpets have been found feeding on one eelgrass plant. The herbivores graze the leaves' epidermis, which contains the chloroplasts with their chlorophyll and plant's photosynthetic machinery. Thus, even though the limpets remove less than 10 percent of the leaf volume, they remove more than 90 percent of the leaf chlorophyll. This cripples the plant's ability to photosynthesize, to manufacture sugar and translocate it to the roots, and to meet its metabolic needs. Soon the plants show stress, the leaves lose buoyancy, growth stops, and the plants die (Zimmerman, Kohrs, and Alberte 1996).

The limpet's graze marks in this photo show the extensive loss of green chlorophyll. Patches of a coralline red alga grow on the limpet's shell.

Tectura depicta's reestablishment in central California may be more evidence of a northward shift of many faunal ranges, associated with a warming trend in our coastal waters.

Polinices lewisii
Phylum Mollusca / Class Gastropoda

Opposite top: Moon snail, about 30 mm long. Eelgrass bed near Monterey's Municipal Wharf II, 10 m deep.

Moon snails spend their time plowing about either at the surface of the sand or just beneath the surface, in search of prey, principally clams but also other snails. Cilia, strong muscles, and quantities of mucus on the foot move this big animal along. It uses its large foot to probe for food. Then, while *Polinices* feeds, its foot holds the victim in place and even envelops and suffocates it (Reid and Gustafson 1989). Note the empty olive and clam shells amidst the debris.

Aided by acidic secretions, a moon snail uses its radula to drill a neat, countersunk hole through its victim's shell to gain access to the interior tissues. *Polinices* then inserts its proboscis and sucks out the soft tissue. This predator does not inject any anesthetic agent or proteolytic enzyme during its attack, though of course the latter enzymes come into play in gastric digestion (Reid and Gustafson 1989). The snail can consume clams without drilling if it finds a gap between their valves.

Today we see few moon snails during our explo-

rations, but it was not always so. In the late 1950s, sea otters had returned to Carmel Bay but had not yet invaded Monterey Bay. SCUBA and neoprene wet suits were new and rare on the scene, and the few early devotees of the sport glimpsed a world that soon was to radically change. Diver George Fraley reports that at that time the sandy bottom near Monterey's Wharf II was covered with thousands of moon snails the size of grapefruits (pers. comm.). For a time George took the snails commercially for sale to Galatin's Restaurant in Monterey. Moon snails are tough, but at the restaurant they were passed through a meat grinder, garlic and spices added, and replaced in the shell. Seasonally, moon snails acquire the toxin that causes paralytic shellfish poisoning, probably by preying on contaminated clams (Carefoot 1977). We hope the restaurant observed the seasonal quarantine not only for those clams but for the moon snails that ate them. The sea otter and the sunflower seastar, *Pycnopodia helianthoides,* are among the moon snails' few predators now that restaurateurs have gone on to serve other fare.

"Sand collar" of a moon snail, 6 cm across. Monterey Bay, 10 m deep.

Moon snails copulate, after which the female extrudes her eggs in a sheet between layers of mucus and sand that she molds about her shell to form a rubbery sand collar strikingly like a disconnected "plumber's friend." Embryos undergo most of their development within the sand collar, so larval planktonic existence is brief.

Neoclinus uninotatus
Phylum Chordata / Subphylum Vertebrata / Class Osteichthyes

Onespot fringehead, about 18 cm long. Monterey Bay, 10 m deep.

On the sandy seafloor adjacent to Monterey's Wharf II we find many bottles and other jettisoned debris. Divers annually clean up this mess, but inevitably it reaccumulates, providing refuge for many creatures, at least for a while. Among these short-term residents are little fringeheads: if the bottles are there, so are these fish, sticking their heads out to see what is going on. In this case, the bottle is covered with red algae. Female fringeheads often lay their eggs in a bottle. The male then takes over, guarding them and ventilating them with his tail until they hatch.

Nobody has described the fringeheads better than Robin Milton Love (1991): "With their funky coloring, little doodily whoppers above the eyes, strange body shapes and great, toothy mouths, these fishes closely resemble descriptions of the aliens."

Citharichthys sp.
Phylum Chordata / Subphylum Vertebrata / Class Osteichthyes

Sanddab, about 40 mm long, on a bat star, *Asterina miniata*. Monterey Bay, 13 m deep.

This little flatfish's camouflage is so thorough we could not have spotted it had it been resting as usual on the sandy seafloor instead of on a bat star. Sanddabs, like several other fishes we discuss, can adjust their colors to match their background by dispersing or aggregating pigment granules in their chromatophores. This little sanddab's color change is a response principally to its visual perception of the surrounding sand, probably mediated by the fish's autonomic nervous system. But information also must be accruing through its left eye that sees the red bat star. Will the sanddab turn red? In addition to such "physiological" color changes, slow color changes in the pigment granules themselves produce "morphological" color changes, due at least in part to hormones. Finally, "structural" colors are produced by light reflected from tiny, mirrorlike surfaces on the fish's body.

During their development, flatfish become adapted for life in a horizontal rather than vertical position. Their eggs are pelagic; early larval flatfish are transparent, and they have a swim bladder so are buoyant enough to live at the surface of the sea. At this stage the flatfish is quite symmetrical, with one eye on each side of its head. As it grows, it descends to the bottom and turns onto one side while its now-lower eye migrates around the head until both eyes are on top, leaving the lower side of the fish blind. The fish's upper side by now has gained pigmentation from its chromatophores. The side against the substratum stays white; without light, its chromatophores do not develop. This shading is adaptive. While resting on the bottom, the fish's upper dark side blends with the substratum; while it is swimming, the fish's light belly does not stand out from below against the surface's bright background. One species of flatfish, a flounder, was kept in an aquarium with blackened top and sides and illumination from below. Dark chromatophores developed in the areas of the fish normally lacking pigment cells as this flounder was induced to prepare itself for an upside-down world (Norman 1975).

In addition to being blind on one side, adult flatfish have lost their swim bladder and so no longer can adjust their buoyancy; when they stop swimming they immediately settle to the bottom again. The sanddab shown here is left-eyed, but some other species of flatfish are predominantly right-eyed.

Pachycerianthus fimbriatus

Phylum Cnidaria / Class Anthozoa

Above right: Tube-dwelling anemone, about 12 cm high. Monterey Bay, 17 m deep.

These anemones live in dense fields scattered across the sandy plain at a rather uniform depth, here surrounded by bat stars and a few skimpy red algae that are clinging to the unstable substratum. A tube-dwelling existence admirably adapts the cnidarians to life in shifting sand, providing both a refuge and stability. They construct their black papery parchment tubes of discharged, interwoven, specialized nematocyst threads that are about 2 mm long and lack spines and hollow tubules (Mariscal, Conklin, and Bigger 1977). So deeply do they burrow that their tubes may attain a length of two meters, most of it buried beneath the sand (Ricketts, Calvin, and Hedgpeth 1985).

Pachycerianthus has two sets of tentacles: longer marginal ones, and shorter inner ones clustered around the mouth. Tentacle color ranges from pale white to orange, yellow, or black. Orange tentacles have a striking, warm, fluorescent glow, apparently produced by their pigments' response to ultraviolet light (Wicksten 1989). An anal pore at the aboral end distinguishes *Pachycerianthus* from most anemones.

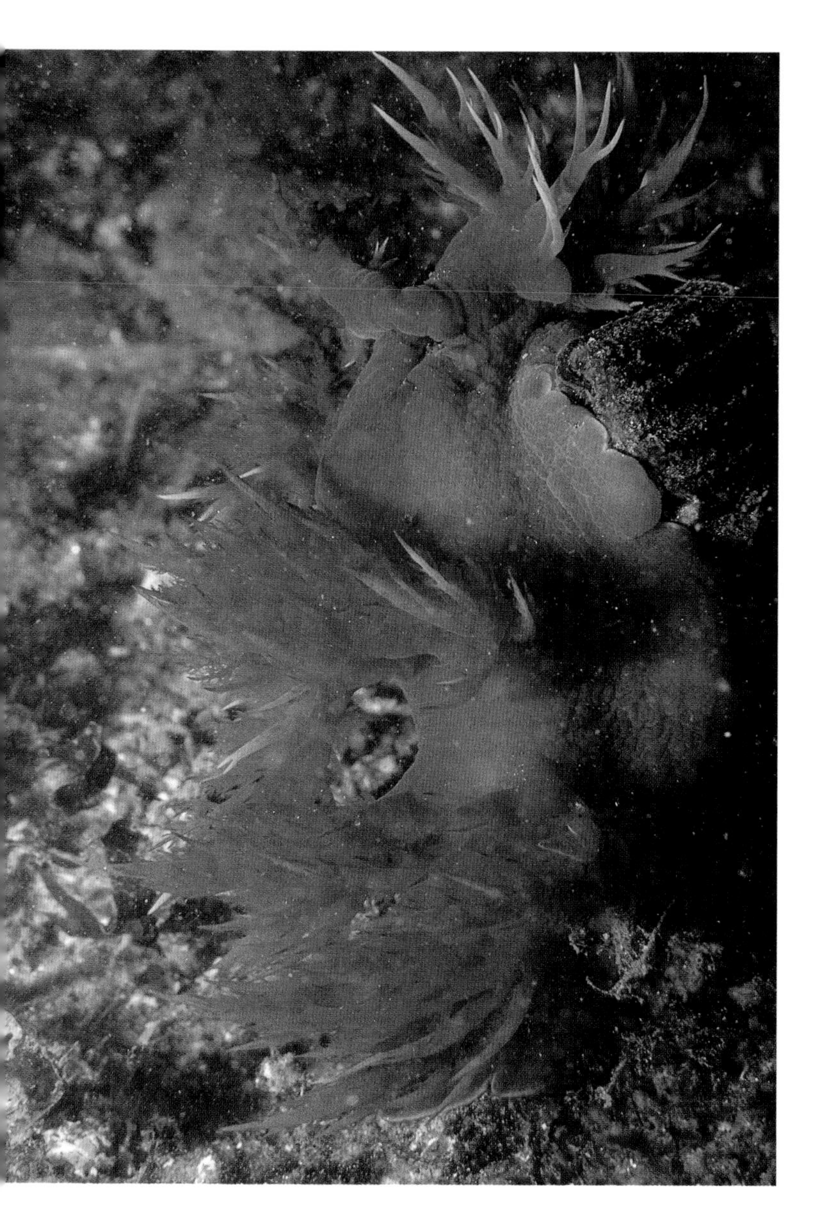

Left: The nudibranch *Dendronotus iris,* about 14 cm long, attacking *Pachycerianthus.* Monterey Bay.

Dendronotus nudibranchs, like so many of their kin, are specialists: they prey only on *Pachycerianthus.* Characteristically, when *Dendronotus* climbs its prey's tube and reaches the extended tentacles, it suddenly lunges forward, seizes a tentacle, and bites it off as the anemone quickly withdraws into its tube, as this one has done. The nudibranch will now proceed to climb into the tube and continue its meal. The anemone makes its defensive movement with longitudinal muscles in its ectoderm, a set of muscles that less specialized anemones do not have. On other occasions, *Dendronotus* seizes a tentacle, gives a violent tug backward, and hangs on so tightly that as the anemone withdraws the nudibranch is drawn head first over the lip and down out of sight into its prey's tube (Wobber 1970)! *Dendronotus* may spend up to two hours in the tube feeding before it reemerges. For the anemone, the loss of a few tentacles is not serious; they will regrow. Adding insult to injury, *Dendronotus* commonly attaches its egg strings on the side of its prey's tube.

Urticina columbiana
Phylum Cnidaria / Class Anthozoa

Sand-rose anemone, about 18 cm wide. Monterey Bay, 20 m deep.

We find these anemones in Monterey Bay at moderate depths, scattered several feet apart on the sandy seafloor. Their columns, like those of their relative *Anthopleura artemisia*, are partially buried in the sand, perhaps attached at their bases to submerged rocks. *Urticina columbiana* has been observed gorging itself on dying squid following their spawning (James Watanabe, pers. comm.), but we find no other records of their feeding habits.

Urticina columbiana was described by Verrill (1922) from specimens obtained by dredging in Puget Sound, Washington, during the Canadian Arctic Expedition of 1913–1918. They were living buried in sand at 60 m depth. "The body-wall in the type figured is clear bright orange-red, and the very numerous conspicuous, raised, verruciform suckers are light yellow"—features not visible in our photographs. In a footnote, Verrill somewhat bitterly states that he had first described the species in 1861, just before the Civil War, from specimens dredged from Puget Sound during the Northwest Boundary Survey, but that the records "were lost, like so many other things in Washington."

Urticina columbiana oral disc, field of view 6 cm. Monterey Bay.

The radiating lines on the polyp's oral disk are external signs of septa that divide the anemone's saccular gut like so many partial curtains, increasing gastric surfaces for the secretion of digestive enzymes and the absorption of food. During the anemone's growth the gastric septa increase in an orderly way from six initially (in these hexamerously symmetrical polyps) to dozens. They support muscles that run the length of the anemone's body and are where the polyp's gametes mature. The presence of septa, of course, is one of the defining features that separates anthozoans from hydrozoans.

Stylatula elongata

Phylum Cnidaria / Class Anthozoa

Below: Sea pen at night, 16 cm high. Monterey Bay, 17 m deep.

Below a depth of about 17 m outside the Monterey Harbor breakwater, many sea pens poke up out of the sand, about a third of a meter apart. This is a nocturnal species; during the day it retreats beneath the sand, a feature that precludes its display in an aquarium. Sea pens are colonial octocorals (soft corals whose polyps have eight pinnate tentacles). As a colony grows, its original polyp is transformed into a huge bulbous anchor below the surface of the sand and a long shaft—the "pen"—that extends above the sand into the water. Hundreds of tiny feeding polyps erupt from the leaflike structures of the pen, and other small polyps circulate water through the colony to keep it hydraulically inflated. A disturbed colony quickly deflates by expelling its water and withdraws into the sand.

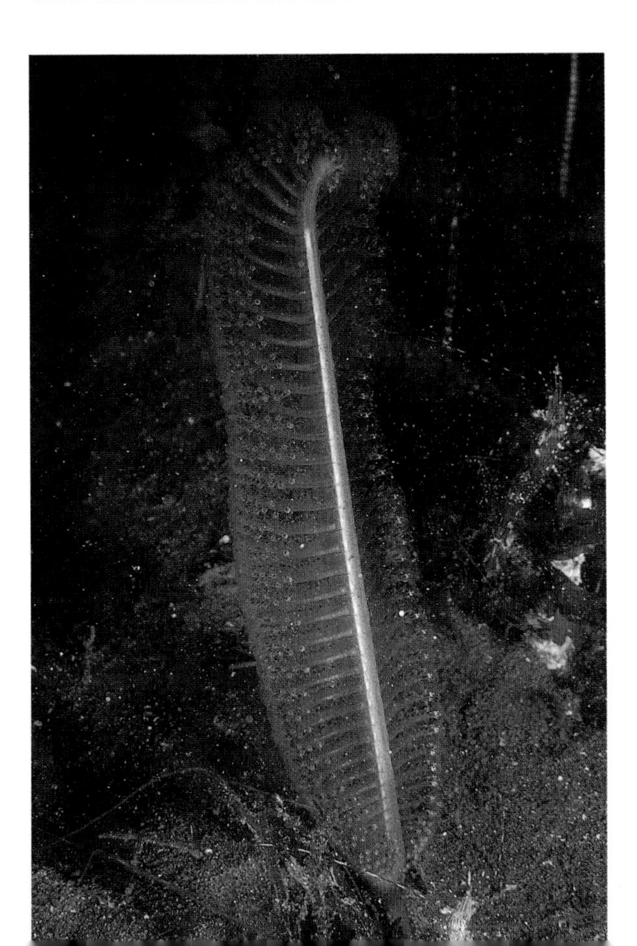

When mechanically stimulated at night, a sea pen exudes a blue or violet bioluminescent slime. The slime contains a photoprotein, which is a stable complex of luciferin, the enzyme luciferase, and oxygen. When calcium activates this system, the photoprotein emits a cold blue light (Pearse et al. 1987). A remarkable diversity of animals, including deep-sea fishes and copepods, squids, polychaete worms, and even fireflies, use the same light-generating chemicals. For some reason, this particular set of components has been hit upon repeatedly in the evolution of luminescent creatures in widely disparate lineages—is it somehow the easiest or least energetically costly mechanism to "invent"?

Luciferin is named for Lucifer, the prince of darkness, the mythical ruler of hell.

Below: Detail of *Stylatula elongata* showing expanded polyps.

These polyps are tightly deployed with almost perfect symmetry. How could a little zooplankter possibly slip through this deadly array of nematocyst-bearing tentacles?

Conus californicus
Phylum Mollusca / Class Gastropoda

Cone snails, about 35 mm. Monterey Bay, 15 m deep.

This is the only species of cone snail, a widespread tropical family, living in California waters. These predatory snails usually live beneath the sand, but occasionally we have seen them emerge to scavenge a dead fish or to join mating aggregations in the spring. Here females are depositing their egg capsules on blades of drift kelp.

Swimming veliger larvae will hatch out in about ten days. Fertilization in these more "evolved" snails (neogastropods) is internal, so complex mating behavior is called for, not just synchronized spawning.

In addition to scavenging, *Conus californicus* feeds on a great variety of living prey including purple olive and black turban snails, several bivalves, a small species of octopus, several polychaetes, an amphipod, and small fish (Kohn 1966). The snail locates its prey by scent, touches the victim-to-be gently with its proboscis, and then suddenly strikes, injecting venom through a large, hollow radular tooth modified into a harpoon. The neurotoxic venom quickly paralyzes the prey, which then is drawn in through the snail's greatly expanded proboscis and swallowed whole.

Some species of cone snails that live on tropical coral reefs are dangerously venomous. One of these, *Conus geographus*, has been responsible for some thirty human deaths. We are told that when the California species strikes it feels like a bee sting (J. Pearse, pers. comm.).

Cancellaria cooperi
Phylum Mollusca / Class Gastropoda

Opposite top: Cooper's nutmeg, 40 mm long. Monterey Bay, 26 m deep.

During the relatively brief bottom time we allowed ourselves at this depth, we were lucky to find this unusual snail, which generally stays hidden in the sand. *Cancellaria* has an implausible prey: bottom-resting Pacific electric rays, *Torpedo californica.* As it seeks a meal, *Cancellaria* leaves trails in the sand which show that it can chemically detect and then turn to approach a resting ray from a distance of 24 meters! Once it reaches the ray, the snail pierces or cuts the fish's ventral skin with its radula and inserts its proboscis. Sometimes it inserts it into the mouth, gill slits, anus, or open wounds of the ray. Then, using its proboscis as a straw, *Cancellaria* quietly sucks blood for almost an hour without eliciting any apparent response from the ray (O'Sullivan, McConnaughey, and Huber 1987). The snail may apply an anesthetic/anticoagulant during its attack, as vampire bats do on land. In the field, up to seven snails have been seen feeding on one quiet ray. In the laboratory, these snails also suck the blood of another cartilaginous fish, the angel shark, *Squatina californica.*

To test whether *Cancellaria* detects rays chemically or by their electric fields, investigators did a Y-maze experiment. The snail moved toward a stream of water that had washed over a ray even though the "ray water" had been brought to the maze in a bucket, thus isolating the ray (and its electricity) from the snail. *Cancellaria* did not move toward water that had washed over bony fishes common in the habitat. Once again, we see a marine creature deciphering a chemical clue (O'Sullivan, McConnaughey, and Huber 1987).

Cancellaria females lay their fertilized eggs in capsules that they suspend above the sand on long stalks, a behavior apparently unique to this snail family. The long stalk may be an adaptation to the deposition of the eggs on unstable sand. Embryos develop in the capsules and hatch out in about a month as feeding planktonic larvae (Pawlik, O'Sullivan, and Harasewych 1988).

Torpedo californica

Phylum Chordata / Subphylum Vertebrata / Class Chondrichthyes

Right: Pacific electric ray, 70 cm long. Monterey Bay, 20 m deep.

This actively swimming ray, still coated with sand, has just left its usual daytime resting place. It swims principally by sweeping its caudal fin from side to side rather than by gracefully sweeping its "wings" up and down, as do most rays. At rest when half-buried in the sand, it is well situated for its role as a daytime ambusher. But it also is vulnerable to bloodletting by *Cancellaria* snails, as we have just seen.

Divers have observed the nocturnal forays of these rays as the rays swim and drift low over reef-tops in search of fish prey that are quiescent there at night. During an attack the ray lunges forward over the fish, quickly folds its pectoral fins down to envelop it, and subdues it with rapid, powerful electric discharges. Then the ray makes barrel rolls or forward somersaults while further enveloping the prey, working it toward its mouth, and finally swallowing the stunned and quivering fish head first and whole. The entire attack from the initial lunge to ingestion of the fish takes about twenty seconds. To confirm this "shock-

ing behavior," researchers attached anesthetized bait fish by wire to a flash bulb and held it before active rays during their nocturnal prowling. In most instances the rays attacked—and dramatically, the flash bulb ignited. So the rays do not capture fish with their mouths, as do most fish, but rather shock and immobilize them first before swallowing them (Bray and Hixon 1978)!

The ray's electric organs lie in its greatly enlarged pectoral fins, and they occupy almost a quarter of the ray's body mass. The organs consist of vertical columns of transformed muscle cells that lack tendons and so are useless for motion. Another species, *Torpedo nobiliana,* can deliver a shock with up to a kilowatt of power. There are fewer layers of connective tissue (that electrically insulate) on either side of the electric organs than over the rest of the body (Rosenthal 1989).

In addition to its offensive batteries, *Torpedo,* like other sharks, skates, and rays, has sensory organs in its nose (with the lovely name "organs of Lorenzini") that can detect extremely small electric fields emanating from its prey. In experiments, the torpedo ray will sometimes attack an energized electrode (Lowe, Bray, and Nelson 1994). The organs of Lorenzini, together with lateral lines that detect minute pressure changes from underwater movements, may account at least in part for *Torpedo*'s uncanny ability to find its prey at night.

Loligo opalescens

Phylum Mollusca / Class Cephalopoda

Squid boats on a summer night. Monterey Bay. (Photo courtesy of Phil Sammet)

During the summer months, at the age of two to three years, market squid—*Loligo opalescens*—return to Monterey Bay in huge numbers to copulate, deposit egg capsules on the sandy seafloor, and then die. To catch them, these two boats work together in a ballet that is carefully coordinated to the squid's behavior.

Seasonally, almost always in the hours before midnight, the squid copulate and the females lay their egg capsules in great piles on the sand. Then, during the early morning hours, the boats go into action. First the squid are located with sonar or with fathometers, and the "light boat" turns on its powerful banks of lights above the massed animals, drawing them, spent after already having spawned, into shallow water and to the dark beneath the light boat. Next the free ends of the squid net are passed from the purse seiner, the "net boat" (the one without illumination), to a skiff in a fixed position. The net boat then deploys the net around the light boat and the massed squid beneath it. The net boat receives the free ends of the net lines from the skiff and pulls the net's bottom together, gradually closing the purse while bringing it to the boat's side. Meanwhile, the light boat departs. The action that follows is dramatic. Sea lions and harbor seals eagerly enter the net to feast, escaping just before the purse finally closes. Blue sharks, however, also avidly feeding within the net, do not voluntarily leave, and fishermen must seize them by the tail and cast them free. Finally, the squirming squid are sucked into the net boat's hull. From one to twenty tons may be taken in one haul, all in less than an hour. In recent years squid have been Monterey Bay's largest fishery.

These boats are within sight of Monterey's Cannery Row—now without canneries, a reminder of the sardine fishery's historic collapse. In theory, taking the squid *after* they spawn should help keep this fishery sustainable. Let's hope it does.

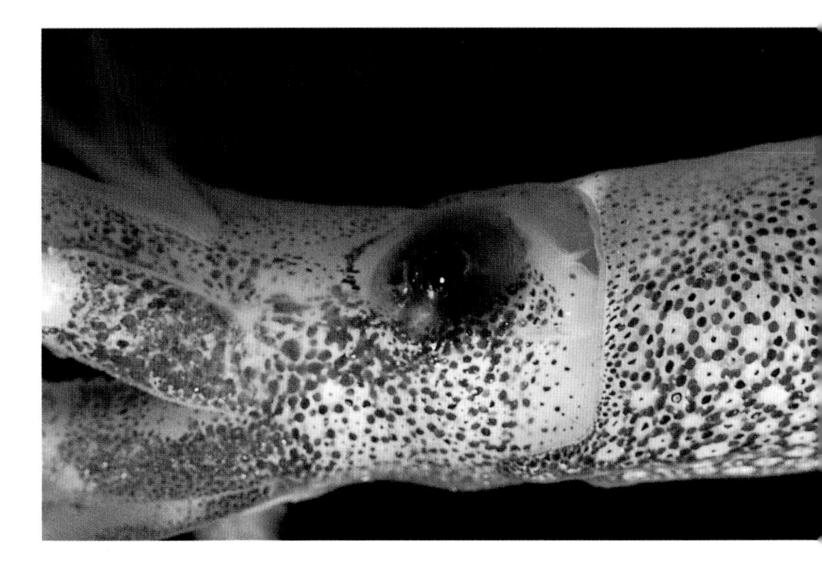

Above right: Market squid, about 18 cm long. Monterey Bay, 20 m deep.

These robust cephalopods, once described tongue-in-cheek as "molluscs trying to be fish," are raptorial carnivores that capture fish, crustaceans, polychaete worms, and occasionally each other. Seabirds prey on them, as do marine mammals, many fishes including salmon and mackerel and bonito, and of course humans. Squid are the fastest swimmers among marine invertebrates. When swimming slowly they use only fin thrust. For an unsustainable, anaerobic, maximum effort of jet propulsion, they contract a belt of muscles to squirt water from their mantle cavity through a funnel. They can reach a speed of eight body lengths per second, attained after two bursts separated by a brief glide while the mantle cavity refills (O'dor 1988). In contrast to most fishes, squid can readily swim either forward or backward; to change direction, they simply rotate their funnel 180 degrees. Catching a fish, *Loligo* abruptly seizes it with its long tentacles, embraces it, then severs its spinal cord with a bite at the back of the head. The squid discards its prey's head and spine while tearing its body to pieces and devouring it (William Gilly, pers. comm.).

Alarmed *Loligo* may escape danger by releasing a black and perhaps distasteful ink cloud that can confuse and even blind pursuers. Besides containing melanin, the ink has alarm pheromones, the neurotransmitters L-dopa and dopamine, that chemically warn other squid of danger (Lucero, Farrington, and Gilly 1994). L-dopa is an intermediate in the synthesis of melanin and dopamine. When diluted ink or L-dopa is applied to a tethered squid's olfactory organ, it elicits a high-pressure jet escape reaction. A local anesthetic applied to the squid's olfactory organ abolishes the response. So even at night or in deep water, when visual cues are limited, squid still can warn each other of danger.

Like other cephalopods, squid exhibit dramatic color changes. Scattered reflective cells embedded deep in the skin produce a brilliant blue-green opalescence. Tens of thousands of chromatophores all over the body create rapid, rippling flashes of red, brown, orange, and yellow. These color changes are under neural control by specific lobes in the brain, driven principally by visual clues. They appear to be related to such activities as feeding, mating, and communication, rather than to any need for camouflage.

The superb, image-forming cephalopod eye has a cornea, iris, lens, and retina, all strikingly similar to the components of a vertebrate eye. Thus, biologists have long regarded the vertebrate and cephalopod eyes as remarkable instances of convergent evolution—analogous (rather than homologous) eyes having an independent origin from rudimentary optical organs in two very different lineages.

For a time, that view was challenged based on genetic evidence. Apparently all animal photoreceptors share a homologous gene (named *Pax-6* in some organisms) that coordinates some part of ocular development in different phyla. Other genes acting downstream may produce the diversity of eye development in different lineages. *Loligo*'s *Pax-6* gene has been cloned and then transplanted to fruit fly wings, antennae, and legs. Fruit fly compound eyes, not cephalopod eyes, sprout at these various sites. The mouse *Pax-6* gene does the same thing: it produces fruit fly eyes at different sites on the insect (Halder, Callaerts, and Gehring 1995; Tomarev et al. 1997).

As we discussed with regard to the compound eye of an amphipod, all photoreceptors also have rhodopsin, a light-sensing molecule that converts the energy of light's photons to electrical energy that can be interpreted by the brain. A comparison of DNA sequences suggests that the protein component of rhodopsin in all animals must share a common ancestry, having been conserved over evolutionary time and tuned to different species (Fernald 1997).

Squid and vertebrates, however, are phylogenetically distant, and despite their eyes' molecular and anatomical similarities, there are also fundamental differences. The vertebrate eye, for one thing, develops as an outgrowth of the brain; the cephalopod eye, in contrast, derives from an ectodermal structure that grows into the brain. Also, the vertebrate retina is inverted: blood vessels and nerves pass in front of the photoreceptors, creating a blind spot at their point of exit. Cephalopods have no blind spot, because the nerves and blood vessels pass behind the photoreceptors. Surely the two designs must have evolved independently.

Summing up the seemingly contradictory evidence, Fernald (1997) states that "homology at the molecular level does not predict homology at the organismic level. The presence of homologous molecules in nonhomologous structures reminds us that molecules are not eyes." J. Gerhart and Kirschner (1997) come to a similar conclusion: "The cephalopod and the vertebrate eye constitute a classic example of convergent evolution on a gross anatomical level, and it is likely that no true conserved eye existed in the ancestor of both lineages."

As so often among animals, exceptional vision goes with uncommon intelligence. Cephalopods like the octopus, whose optic lobes are much larger than the rest of their brain (Hanlon and Messenger 1996), can be quickly conditioned to respond to geometric patterns such as triangles and squares, they can learn to imitate the behavior of others, and they have excellent long-term memory. They have a half billion nerve cells, one-third of them in the brain—more than any other invertebrate, even more than many fishes and most reptiles.

Spawning *Loligo* at night. Monterey Bay, 30 m deep. (Photo courtesy of Phil Sammet)

Here are spawning *Loligo* hovering above a cluster of egg capsules. Two males, whose arms have turned dark red, are embracing females whose arms are clear. The female on the left holds a capsule of fertilized eggs, tilted up in her arms, ready for attachment to the communal egg cluster at the sea bottom.

During copulation, the female's eggs are extruded from her oviduct, gathered into masses of about two hundred, given several coatings of jelly (a glandular secretion), and allowed to harden into spindle-shaped, sheathed capsules. Meanwhile, a male, using a specialized arm, thrusts several sperm packets into her mantle cavity. Once there, the packets' caps blow off, releasing clouds of sperm. Fertilization occurs during the formation of the egg capsules. After she deposits her egg capsules, her mate releases her and both squid swim away, perhaps to mate with others.

But neither will survive for long; reproduction leads squid of both sexes to exhaustion and death. In the extraordinary effort, many are mutilated as grasping suckers strip off skin, tattered arms hang loosely after motor control is lost, animals darken as chromatophore control is lost, some swim erratically or bob vertically at the surface. This is "big bang" reproduction: a short life of one or two years followed by prolific mating, egg-laying, and death. Shearwaters (*Puffinus* sp.), gulls (*Larus* sp.), giant sunflower seastars (*Pycnopodia helianthoides*), marine mammals, and blue sharks will feast on the dead (Fields 1965).

Below: Female *Loligo* depositing an egg capsule at night. Monterey Bay, 30 m deep. (Photo courtesy of Phil Sammet)

The female in this photo is grasping an egg capsule vertically in her arms and blowing away sand before depositing the capsule on the sea bottom. The capsules are expelled through the siphon, apparently with great effort, by successive contractions of the mantle. In aquaria, one female may lay twenty capsules that will swell on contact with seawater and then together weigh more than she does (Fields 1965). Spawning females are attracted to existing egg capsules, and eventually mounds of capsules a meter and a half high will festoon the bottom. After three weeks of embryonic development, juvenile squid will hatch from the capsules, forgoing a larval stage.

Dead squid are quickly eaten or decay, but their egg capsules are largely resistant to predators and

to bacterial and fungal infections (Fields 1965). A symbiotic bacterium can be cultured from the egg capsule sheaths and from females' secretory glands that appears to produce antibiotic and antifungal substances and highly volatile compounds (polyamines); these may help protect the eggs from infection and from animal predators (Kaufman et al. 1998). The MacGinities (1968) fed squid eggs to anemones, and the egg mass was regurgitated a few hours later; the embryos continued their development unharmed by the experience. A solution from the bacterial cultures causes a seastar to retract its tube feet, thus thwarting its means of attack. Some powerful defensive chemical warfare is going on here!

Opposite: Hatchling squid, 4 days old, about 5 mm long. From Monterey Bay Aquarium.

It is fascinating to observe these little babies in the laboratory; they jet forward, they jet backward, and their chromatophores produce starry flashes of yellow, red, and brown over their body. Sometimes—in a kind of molluscan youthful exuberance?—they squirt little blobs of ink.

This is the age when the hatchlings learn hunting skills needed to capture copepods, an erratic prey with quick, evasive reactions but a crucial food for squid this tiny. By trial and error, juvenile squid must learn to encircle the copepod, make a close approach from in front, and suddenly seize it with a "tentacular strike." Pursuit from behind is doomed to fail. They will use these techniques later to capture shrimp and fishes. For squid, the first forty days of life are a critical period for learning. Should the hatchlings during these weeks encounter only slow-moving prey such as brine shrimp, they will never acquire the hunting skills they eventually will need in the wild (Chen et al. 1996). And learn they must or they face starvation; the energy reserves they are born with last only a few days.

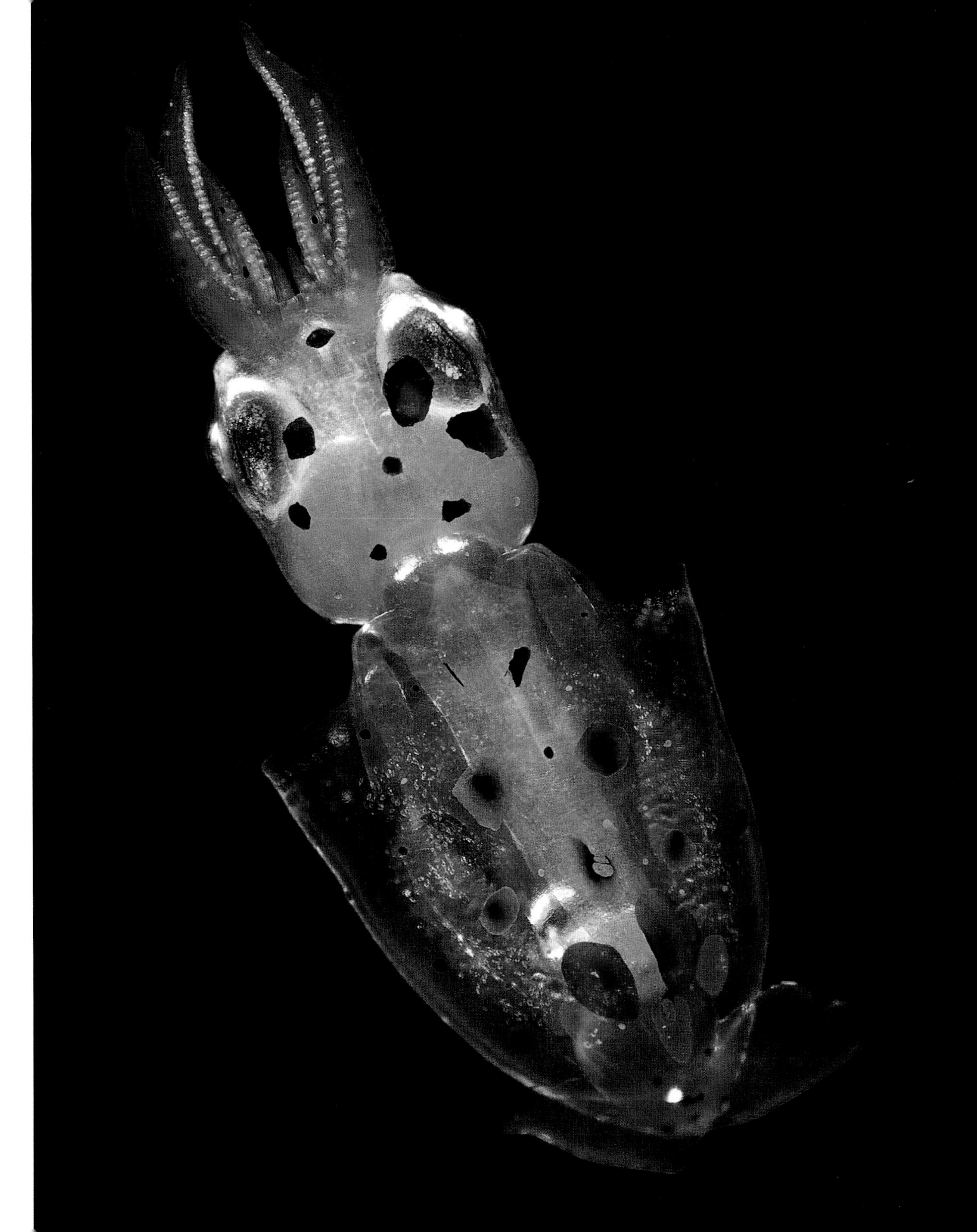

SEA CRANE
MBARI
Ventana

7 The Monterey Canyon

A huge submarine canyon cuts the floor of Monterey Bay, extending from the shore at Moss Landing to the edge of the continental shelf. Within the outer bay it has a depth of 1300 m and a profile almost like that of the Grand Canyon of the Colorado River. It slopes down to a depth of 3200 m at its seaward end. There is indirect evidence (albeit disputed) that this remarkable geologic feature was excavated about 25 million years ago, when the land was higher (or the sea lower), by a river that flowed westward from the Central Valley and emptied into the Pacific Ocean where Santa Barbara now stands. Subsequently, northward movements at a rate of about 4 cm/yr carried the Pacific plate, bearing the future Monterey Bay, about 560 km along the enormously complex San Andreas fault system to its present location (Bergeron 1997). Monterey Canyon cuts through sedimentary rocks that are being eroded and reexcavated by "turbidity currents," underwater currents that carry heavy loads of sediment which scour the rocks they traverse.

Explorations of the canyon in 1990 with the submersible *Alvin*, and subsequently with remotely operated vehicles (ROVs) sponsored by the Monterey Bay Aquarium Research Institute (MBARI), have produced a wealth of information, documenting communities of unusual organisms previously known only from dredges and fishermen's trawls. These organisms are associated with "cold seeps" of hydrogen sulfide and methane gas at depths of 580–3200 m. The gases appear to arise from the anaerobic decomposition of huge deposits of buried microscopic plants and animals such as diatoms and radiolarians. It's an environment of total darkness—there is no photosynthesis, so no algae. The temperature may be a low 5°C, oxygen levels are greatly reduced, and the water pressure can be an immense 318 atmospheres (318 bar or 4,675 psi).

The seeps are dominated by mats of chemoautotrophic bacteria—bacteria that derive energy from inorganic chemical compounds rather than from the sun. Deep-

MBARI's ROV being lowered over the side of its mother ship.

A section of the canyon wall as represented in the Monterey Bay Aquarium. View about 60 cm.

sea clams harbor these bacteria in a symbiotic relationship, with the energy harnessed by the bacteria being used to convert dissolved carbon dioxide into organic compounds the clams can exploit. Other fauna associated with the seeps include worms, crabs, and snails. Over fifty other species of non-seep-dependent benthic fauna have been found so far in the deep canyon and many more are bound to turn up (Barry et al. 1996).

Remotely operated vehicles like the ROV *Ventana* deftly collect specimens from the depths of the canyon without risking human life. Sit in the control room of the mother ship where the ROV and its cameras are being driven and managed, peer at the TV monitors, and you quickly experience the illusion that you are deep down in the ROV itself.

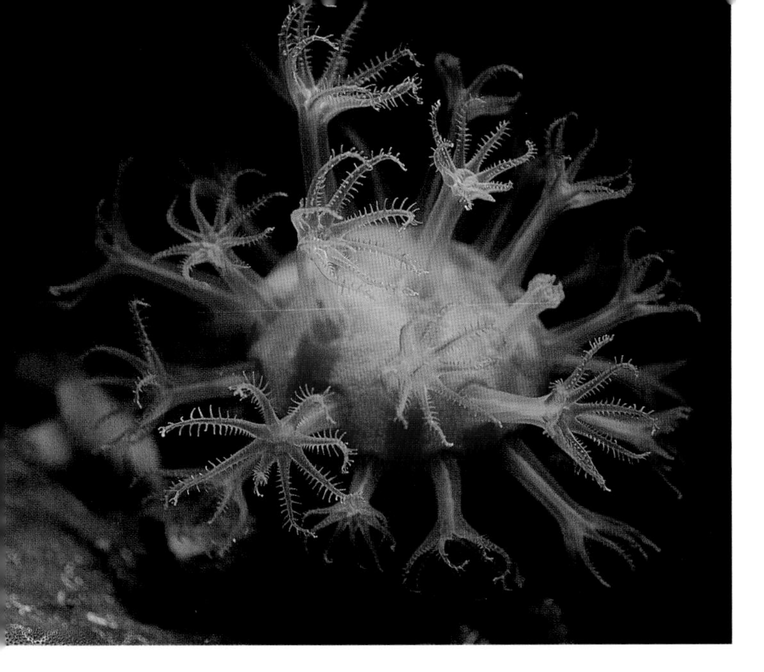

Anthomastus ritteri
Phylum Cnidaria / Class Anthozoa

Mushroom coral, about 5 cm wide. From Monterey Bay, 360 m deep. Laboratory, Monterey Bay Aquarium.

In 1904, the U.S. Fisheries steamer *Albatross* explored California's coastal waters, including Monterey Bay, collecting fauna by the hit-or-miss method of deep trawls. On that memorable cruise, fourteen new species of octocorals (Alcyonaria) were found, among them *Anthomastus ritteri*. One of the naturalists on the *Albatross* described this species as "an early rose potato stuck full of red cloves" (Nutting 1909).

How technology has changed! Now, using MBARI'S deep-diving ROVs, scientists carefully select and collect these specimens, leaving the rest of the habitat unscathed.

Anthomastus lives on rocky walls and on sediment-covered habitats at depths ranging from 360 to 1200 m. This soft coral has two kinds of polyps: tall, slender feeding polyps and polyps, appearing as small bumps embedded in the "mushroom," that circulate water through the colony (Abbott 1987). The tentacles' nematocysts are sticky rather than venomous. Fecal pellets from freshly collected specimens consist of sand and particles in which sponge spicules, small eggs, foraminiferans, and diatom skeletons can be identified. The polyps apparently bend over to feed on settled material.

Note the eight pinnate tentacles of each polyp, characteristic of octocorals; the esophagus of each polyp is orange. In the laboratory, the coral is unable to capture larger, active organisms like mysids, which break free from the polyps' sticky tentacles, but it readily takes live brine shrimp larvae or dead krill (Riise 1990). A "primary tentacle" moves small captured food particles to the mouth; for larger particles, the other tentacles participate by bending over to form a cage. Peristaltic waves move the particles down the esophagus into a system of interconnected tubes.

Many *Anthomastus* corals in the canyon are relatively large, possibly 18 cm across. The lower portion of their column may be covered with sediments; polyps apparently cleanse the upper portion of the column as they feed. The colonies' colors range from white to pink or even red. They reproduce with crawl-away larvae.

Anthomastus with retracted polyps, about 30 mm wide. Laboratory, Monterey Bay Aquarium.

Anthomastus exhibits a rhythmic pattern of activity, retracting and extending its feeding zooids in a repeating cycle of 16–38 hours—a rate that varies for each colony and is apparently not influenced by environmental cues such as light or food availability (Chandrasekaran 1991). Because the cycles of different colonies are not synchronized, expanded and contracted colonies may lie adjacent to one another. The polyps remain retracted for 2 to 5 hours before reexpanding. But physical disturbance, particularly of the column, will cause the polyps to retract even during their feeding cycle. This could be a defensive reaction to animals that prey on octocorals such as the nudibranch *Tritonia diomedea*, which often lurks nearby (Riise 1990).

At its ambient temperature of 7°C, *Anthomastus* seems to do everything in slow motion. It may take almost an hour for a polyp to capture a large food particle and move it down through its esophagus into a system of interconnected canals. Growth of colonies in the laboratory is very slow, leading to speculation that older colonies in the canyon may be up to one hundred years old.

Umbellula lindahli
Phylum Cnidaria / Class Anthozoa

Droopy sea pen, about 14 cm tall. Laboratory, Monterey Bay Aquarium.

These sea pens anchor themselves in soft sediments at depths below 500 m in the oxygen minimum zone throughout the Monterey Bay Canyon. The species first was discovered from a whaling ship in Arctic waters in 1753, and the generic name was assigned by Cuvier ("father of comparative anatomy") in 1797. Subsequently, the strange, deep-dwelling animals have been found in all the world's oceans. One was photographed at a depth of over 5000 m on the abyssal plains west of South Africa; it looked like a pinwheel on a stalk and was estimated to be a meter high—similar to this one but much taller. *Umbellula* also looks strikingly like the stalked echinoderms known as crinoids and a stalked bryozoan, organisms from different phyla that live under the same conditions in the deep sea—suggested as examples of convergent evolution (G. Williams 1997).

These sea pens, like their cousins in shallower waters (see chapter 6), are colonial octocorals. Their penlike primary polyp gives rise to feeding polyps and to the anchoring foot; other polyps circulate water through the colony. *Umbellula* has only a few very large tubular feeding polyps that radiate from a central point, each with an octocoral's characteristic eight pinnate tentacles. It's flexible internal "pen" raises it up into the prevailing currents, which often are strong enough to bend the colony toward the bottom, hence its common name. These currents sweep nutrient particles along the deep-sea bottom, where *Umbellula*'s outstretched polyps are well positioned to catch them.

Umbellula's pen has a muscular "foot" that pushes a few millimeters into the sediment. The foot secretes a mucus collar that cements the sand around its base. In the laboratory, from time to time these sea pens detach themselves from the sediment with peristaltic waves of the foot, drift to a new location, and then reattach themselves. The colony leaves its mucus collar in the sediment when it changes its position. As in so many deep-sea organisms, mechanical stimulation causes *Umbellula* to bioluminesce, but these colonies are special in that they emit sparkles of both blue and green wavelengths along different parts of their bodies (Widder, Latz, and Case 1983).

Captive *Umbellula* can be maintained in seawater at 5–6°C and an oxygen concentration of 0.7 ml/L, conditions that prevail at the collection site. The greatly reduced oxygen concentration is obtained by bubbling nitrogen through the water. Maintaining *Umbellula* is a labor-intensive task: one must feed each polyp individually, for rather than delivering harvested bits of food to a central body cavity for digestion, as in most cnidarians, *Umbellula* polyps process their food themselves (Jason Flores, pers. comm.).

Laqueus californianus var. *vancouveriensis*
Phylum Brachiopoda / Class Articulata

Articulated brachiopod (lamp shell), 30 mm across. From walls of Monterey Bay Canyon. Laboratory, Hopkins Marine Station.

Brachiopods are odd creatures, lophophorates related to bryozoans and phoronids but in a phylum all their own. These individuals live clustered on the canyon walls at depths of 500–700 m, where it is dark, cold (8–12°C), and under enormous pressure from the overlying seawater. Brachiopods' calcium carbonate shells superficially resemble those of bivalve molluscs, but the two valves of a brachiopod lie above and below the body, not to the body's left and right as they do in molluscs. In this brachiopod species the valves are hinged, but in some (the so-called inarticulates) they are joined by muscles alone. The ventral valve bears a short anchoring stem, while the dorsal one supports a big lophophore (the coiled tentacle-bearing feeding and respiratory organ), which lies protected within the capacious mantle cavity. In life the animal flops over "on its back" so that its stemmed valve lies atop its dorsal valve, as seen in the photo. Beating cilia on the lophophore draw water through the mantle cavity, providing oxygen and food and carrying away dissolved wastes. *Laqueus* has no anus: it regurgitates fecal wastes and then ejects them periodically by snapping its shell.

Brachiopods left a rich fossil record, with more than 12,000 known species in the Paleozoic Era, but they were almost wiped out during the great Permian extinctions 245 million years ago. Only about 350 species of brachiopods exist today, most of them articulates. A few species are fairly common where they do occur.

Mussels generally dominate the shallow habitats where both they and brachiopods might be expected to live. When they are placed in proximity experimentally, mussels rapidly smother and outcompete brachiopods. As prey, brachiopods seemingly have little to fear. Their flesh appears to be repellent; it causes fish to convulse and regurgitate, and snail and seastar predators preferentially take mussels before brachiopods (Thayer 1985). Also, brachiopod tissues are of a very low density and contain inorganic spicules, diminishing their nutrient value to predators (Peck 1993).

Sexes are separate. Although many brachiopods brood, *Laqueus* appears to be free-spawning. The larvae of the closely related *L. californianus,* at least, settle preferentially on the ventral valves of living *Laqueus* brachiopods, a behavior that may account for the dense clustering of these animals. Valves of dead *Laqueus* induce settlement much less, and other materials have little or no effect on the larvae (Pennington, Tamburri, and Barry 1997).

Opposite: Dorsal valve with lophophore, 34 mm across. From laboratory, Hopkins Marine Station.

Removal of the ventral valve exposes the thin mantle lining the valves of *Laqueus* and the elaborate respiratory and feeding lophophore. A coiled cartilaginous tube supports ciliated filaments that are unattached at one end, as seen here. Comparable filaments in the gills of bivalve molluscs, in contrast, are "tied down" at both ends. Thus brachiopods must shield their lophophores against too strong a current, lest their respiratory and feeding filaments be thrown into disarray. Ciliary tracts move water over and between the filaments and pass it out anteriorly (to the right in this photo) through a gap between the valves. Meanwhile, food particles are trapped and moved basally along the filaments to a food groove and on to the mouth. *Laqueus* appears to feed on detritus, organic silts, and dissolved organic matter.

Laqueus moves water through its mantle cavity and lophophore in a strictly laminar flow pattern with no mixing of filtered and incurrent water. It's a low-energy system with a filtration rate only one-third to one-sixth that of mussels. This means that brachiopods move water more efficiently than do bivalves; the latter sacrifice energy efficiency for much higher flow rates. This low-energy system correlates with the brachiopods' extremely low metabolic rate. Indeed, they may have the lowest oxygen consumption per unit of weight of any animal on the planet (LaBarbera 1981; Hivnor 1986). LaBarbera suggests that in Paleozoic times, when brachiopods first evolved, the sea's productivity may have been very low; because there wasn't much food around for them to feed on, it was critical that they extract every possible morsel available. Perhaps the lophophore's inherent design prevented brachiopods from imitating bivalves once food was more plentiful.

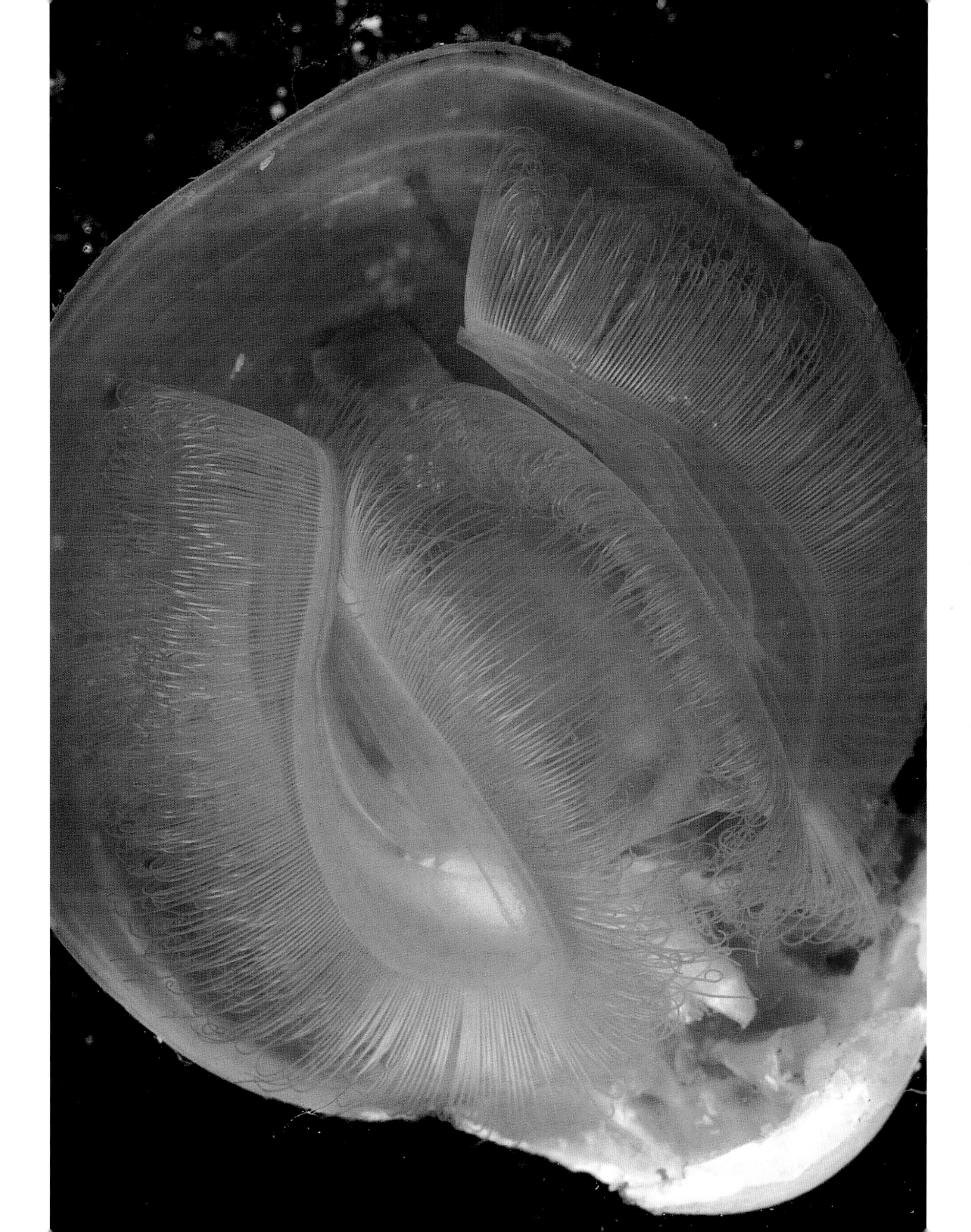

Megalodicopia hians
Phylum Chordata / Subphylum Tunicata / Class Ascidiacea

Predatory tunicate, 9 cm high. From 305 m, Monterey Bay. Laboratory, Monterey Bay Aquarium.

This ascidian tunicate, widespread in deep water along the West Coast, lives in patchy but often rather dense groups where rock and mud are intermixed. Standing up to 15 cm high, it is big for an ascidian. And unlike its filter-feeding relatives, this extraordinary species (and other members of its abyssal family, the Octacnemidae) engulfs big prey—mostly larvae and crustaceans as big as mysids—that venture near its gaping muscular mouth-lobes (Monniot, Monniot, and Laboute 1991). The enormous mouth cavity leads to a "throat" in back, and thence to a rather short gut. The whitish mass of the gonads lies alongside the gut. The pharynx, which in more familiar ascidians is an elaborate food-trapping basket, is greatly reduced and simplified in octacnemids. The gills are merely a few holes that conduct a weak respiratory current from the pharynx into the cloacal cavity. Mucus, which in more familiar ascidians makes nets to trap micro-food drawn with the respiratory current into the pharynx, here forms a rope that is extruded from the pharynx into the mouth cavity to entangle prey there. The cloacal siphon is very small compared to the mouth and sticks up like a little chimney behind the oral lobes.

Like most abyssal organisms, even after it has been found this predatory tunicate is difficult to retrieve. The ROV's pilot uses a clamshell arm to break off the substratum with the tunicate attached. Then the specimen is brought up in an insulated drawer in the submersible. The animal is unlikely to survive even a slight scrape or tear.

Megalodicopia has lived for over two years in chilled, low-oxygen aquaria at the Monterey Bay Aquarium, which afford observations of its behavior. The tunicates detect vibrations, for example those of a swimming crustacean, but they are unaffected by light. Physiological processes proceed slowly. Food from weekly feedings takes well over a week to traverse the gut and be ejected as waste from the cloacal siphon. The internally fertilized eggs become sticky soon after release and adhere to nearby hard substrata; there they develop slowly into tailless larvae, which have not been induced to metamorphose in the laboratory (Gilbert Van Dykhuizen, pers. comm.).

Florometra serratissima
Phylum Echinodermata / Class Crinoidea

Opposite: Feather star, about 14 cm long, on peppermint gorgonian coral, *Paragorgia* sp. Laboratory, Monterey Bay Aquarium.

Adult *Florometra* are surprisingly agile, climbing about freely using their arms to attain an exposed position. Feather stars are suspension feeders. In contrast to seastars and most modern echinoderms,

their mouths usually face upward, and when feeding they extend their arms to form a funnel to gather in suspended detritus. This *Florometra*, in typical crinoid fashion, has climbed onto a gorgonian coral that it clutches with its leglike cirri. Uncharacteristically, however, it faces downward, and it has thrived for months in this position. A brisk intake flow on the opposite side of the aquarium is downwardly directed, bringing an upwelled current—and any introduced food particles—toward *Florometra*'s extended arms, thus explaining the feather star's anomalous position.

In Monterey Bay, *Florometra* ranges in depth from about 30 m to over 1000 m and thus into the oxygen minimum zone. They live densely clustered, perhaps as many as one hundred per square meter, on rocky ridges and on the edge of the slope of the Monterey Canyon. To escape predators, *Florometra* swims away with coordinated flapping motions of its feathers (Shaw and Fontaine 1990).

Note the peppermint gorgonian's tiny, extended feeding polyps, each bearing eight tentacles. On coral reefs, we have observed that gorgonians extend their polyps into a flowing current and retract them at ebb tide. In its laboratory setting, this peppermint gorgonian behaves similarly: it retracts its polyps when the intake flow is turned off.

Opposite: Feather star, about 12 cm across. From Carmel Bay, 58 m deep. (Specimen courtesy of Mike Guardino)

Florometra has ten arms, each bearing rows of "pinnules" lined by tiny tube feet, which on some pinnules are barely visible as a fringe. The upwardly directed mouth apparently is closed in this view, but the anus is visible at the end of a short nipple. During feeding, suspended particles that come in contact with the tube feet are trapped in mucus and flipped into a ciliated groove along which they are moved toward the mouth.

In *Florometra* the sexes are separate. Fertilization is external, and swimming, nonfeeding larvae settle gregariously, often next to or on adults. Developing feather stars progress through a stalked form reminiscent of their ancestors. These "pentacrinoids" grow in clusters, their stalks arising from tightly packed attachment sites, like a bouquet of lilies spreading from a vase. At about six months of age, pentacrinoids metamorphose into juveniles by detaching from their stalk and developing pinnules and cirri (Mladenov and Chia 1983).

Crinoids are the first of the modern echinoderm classes to appear abundantly in the fossil record. Whereas the sea lilies, whose arms and central bodies cluster atop long, skinny stalks, date back to early Paleozoic times, 500 million years ago, modern crinoids first appeared in Triassic times, 250 million years ago. A few of these species are stalked, but most are as mobile as the adult *Florometra*.

Calyptogena pacifica

Phylum Mollusca / Class Bivalvia

Cold seep clams from 700 m depth, Monterey Bay Canyon. Laboratory, Hopkins Marine Station.

In 1988, during dives deep into the Monterey Bay Canyon, researchers using the submersible *Alvin* discovered "cold seeps" and associated communities of organisms living symbiotically with bacteria. At depths of 580–3200 m in the seaward mouth of the canyon and on its walls, hydrogen sulfide rises from scattered fissures (seeps) in shales and fractured sandstone. The clams that live there may look like your everyday clams of the mudflats, but they lead a very strange life. At a depth of 3200 m, the pressure is 318 atmospheres (318 bar or 4,675 psi), the temperature a chilling 5°C, and the oxygen concentration only a tenth that of surface waters. The hydrogen sulfide accumulating here irreversibly poisons hemoglobin in an animal's red cells and binds to enzymes that catalyze aerobic metabolism, so it's highly toxic to most air-breathing organisms. As for food, at this depth the clams are in constant darkness, far below the blooms of the photosynthesizing phytoplankton that might feed them. In such an unfriendly environment, what is a clam to do?

Calyptogena pacifica meets its nutritional needs by exploiting huge numbers of symbiotic intracellular bacteria that it harbors in its gills. These bacteria, which are passed from one generation of clams to the next via the eggs and larvae (Cary and Giovannoni 1993; Lisin, Barry, and Harrold 1993), make organic products from dissolved carbon dioxide and seawater in a two-step process called "bacterial chemosynthesis." First the bacteria oxidize ("burn") hydrogen sulfide to get the energy needed to synthesize adenosine triphosphate (ATP), the same energy source that plants produce in photosynthesis. Spending ATP, the bacteria next use dissolved carbon dioxide and seawater to make organic compounds such as sugars: food for the clam. The processes are analogous to photosynthesis in green leaves. While plants use sunlight for energy, the bacteria use an inorganic fuel, hydrogen sulfide. A schematic equation describes the events: $CO_2 + H_2S + O_2 + H_2O \rightarrow CH_2O$ (carbohydrate) + H_2SO_4 (Jannasch and Mottl 1985).

The hydrogen sulfide on which these bacteria depend diffuses on either side of the fissures. *Calyptogena*'s feet are buried deeply in the sulfide-rich sediments there, and their siphons extend into the water. Tracks show that the clams congregate about two meters from the seeps, where they evidently find the proper H_2S concentration. If the H_2S emission from a fissure slows or ceases, the clams move on to more productive sites. Note that in these specimens the corrosive H_2S has even eaten away the shell's proteinaceous outer covering, leaving the white calcareous layers exposed.

The H_2S binds to a specialized zinc-containing protein in the blood serum of the clam's vascularized foot. This protects the clam's hemoglobin and enzymes from sulfide toxicity. Made innocuous this way, the H_2S is brought by the blood to the large, fleshy, red gills. At the same time, the clam draws seawater over its gills, bringing oxygen to the bacteria lodged there. Thus the necessary ingredients—H_2S for energy, oxygen for respiration, and specialized bacteria for chemosynthesis—converge for the symbiosis to work its magic.

Cold seeps (and hydrothermal vents) are ephemeral sources of hydrogen sulfide; they often arise at widely separated deep-sea sites only to exhaust themselves in decades, centuries, or perhaps millennia. So how do the associated fauna come to populate these often distant sources of chemical energy? In 1988, during a dive by *Alvin,* the remains of a large baleen whale were discovered at 1240 m depth in the Santa Catalina Basin, remote from any cold seep. Thick bacterial mats similar to those at cold seeps covered its exposed skeleton, and many of the associated invertebrate fauna, including clams, were species also found at cold seeps that derive their energy from symbiotic chemoautotrophic bacteria. It seems that anaerobic bacteria were decomposing the lipid-rich whale bones and releasing hydrogen sulfide, which supported both free-living and symbiotic bacteria that in turn provided the "whale fall" fauna with its nutritional needs. Many such whale falls have subsequently been discovered. It may be that these nutritional oases are stepping stones for larvae in their dispersal between cold seeps or hydrothermal vents (Bennett et al. 1994).

Sebastolobus altivelis
Phylum Chordata / Subphylum Vertebrata / Class Osteichthyes

Longspine thornyhead, 22 cm long. Deep sea laboratory, Monterey Bay Aquarium.

These bright red bottom-dwelling rockfish live at 400 to 1200 meters depth on the slopes of the Monterey Bay Canyon. Despite their great depth, they, as well as shortspine thornyhead, sablefish, and Dover sole, increasingly are being taken commercially by bottom trawls and by longline gear. All these fish are very slow growing, slow to reproduce, and long lived; the longspine thornyhead, for example, has a life span up to forty-five years. Such traits render them vulnerable to overexploitation even by a modest fishery (Jacobson and Vetter 1996).

In the spring, *Sebastolobus* females spawn buoyant gelatinous egg masses that float to the surface, where both the egg masses and newly hatched feeding larvae have been recovered. Juveniles live at midwater depths for eighteen to twenty months before settling to the bottom, and can be so numerous that in March during daylight hours they form part of the "deep scattering layer" at 600 m depth. One advantage of this up-and-down life cycle is that as larvae and juveniles (which feed largely on krill) they have the benefit of rich food resources in relatively shallow water before their descent to a benthic life. Of course, they are also subject to heavier predation (K. Smith and Brown 1983; Wakefield and Smith 1990).

The adult longspine thornyhead, a bottom dweller that feeds largely on brittle stars, lives and reproduces in the oxygen minimum zone. In addition to tolerating reduced oxygen concentrations, these fish must cope with constant low temperature, a very high hydrostatic pressure, meager light, and a scanty food supply. One of the thornyhead's adaptations to this environment is a low metabolic rate. They are so sluggish that they can be approached within a few centimeters by the submersible ROV's manipulator before they swim a short distance away and come to rest again (K. Smith and Brown 1983). Certainly they don't need speed to capture brittle stars, and they may not be subject to much predation at depth. Correlated with a sedentary lifestyle is a very low concentration of the enzymes (dehydrogenases) in their white skeletal muscle that break down glucose anaerobically (Sullivan and Somero 1980). These muscles are used only for brief bursts of swimming; they have a reduced blood supply and utilize predominantly anaerobic metabolism.

Another adaptation of *Sebastolobus* lies in the way many essential enzymes are altered by pressure over the fish's life span. For example, the dehydrogenases that mediate muscle contraction become irreversibly deformed and pressure-dependent when the juveniles descend to the bottom. Thus adult *Sebastolobus* do not survive when brought to the surface; the fish depicted here was caught in midwater as a juvenile and reared to adulthood in an aquarium. In contrast, the same enzymes of shallow-water fishes, when changed by the high pressure at great depth, simply cease to function. Shallow-water fishes therefore do not survive when taken into deep water (Siebenaller 1984).

In addition to being caught as a juvenile before its proteins were irreversibly altered by pressure, this fish had its swim bladder bled with a syringe and tiny needle before being brought to the surface to be reared in an aquarium (Gilbert Van Dykhuizen, pers. comm.). Without this intervention, the swim bladder's expanding volume of air due to pressure reduction at the sea surface would have caused this juvenile almost to explode. What a sophisticated effort it took to bring this fish successfully into an aquarium!

Sebastolobus and several other fishes, together with a few seabirds, shorebirds, and marine mammals, are among the few vertebrates that concern us in this book. They do feel like our relatives in nature; they grab our attention and imagination in a way that other organisms simply do not. Perhaps we have a cousin's sense, however inchoate, of a bird's or a fish's predicament in a way we simply don't of a sponge's or a snail's. And yet here we present a book that deals largely with invertebrates and seaweeds. We sincerely hope that your learning experience has mirrored ours and that the kelps, the sponges, the snails command a newfound sense of excitement and respect for their beauty and their sophistication.

APPENDIX
An Introduction to the Animal Phyla

In this book we call animals and seaweeds by unfamiliar, italicized names, and we group them into sometimes oddly named "phyla" and classes. Most marine animals lack common names beyond ones that refer to their general groups—seastar, hermit crab, jellyfish, sea squirt . . . But each described alga, animal, and plant carries a unique Latinized name, bestowed by its describer and used worldwide. Ever since Linnaeus established modern taxonomy in the early eighteenth century, each kind of organism has been given a double name—its genus first and then its species. For over 250 years, scientists have devised and agreed upon the names of every known organism on earth, living or fossil, according to this Linnean system of nomenclature—a remarkable political record as well as a stunning intellectual accomplishment.

The species is the narrowest formal taxonomic unit to which an organism belongs. Just about the broadest is the phylum, of which there are some thirty-five currently described in the animal kingdom. Phyla are based on distinct, recognizable structural designs. They are in this inclusive sense "basic kinds" of organisms. In between are the successively narrower classes, orders, families, and genera.

In the appendix below, we cite examples in the book representative of the various phyla, so you can meet these animal types pictorially—setting up, we hope, future meetings in the flesh. Then we review some of their distinguishing (or simply interesting) traits, so you may know in a more biographical way what our photographs depict: the biology behind appearances, so to speak.

This appendix summarizes only the most prominent phyla, and then in brief. Especially in the case of the invertebrates covered here, we leave detailed characterizations to such excellent invertebrate zoology textbooks as Morris, Abbott, and Haderlie 1980; Nybakken 1997; Barnes 1980; Brusca and Brusca 1990; and Pearse, Buchsbaum, and Buchsbaum 1987.

PHYLUM PORIFERA (Gk. *poros*, channel; L. *ferre*, to bear)

For Example:

Calcareous sponges—Class Calcarea—p. 136
Demosponges—Class Demospongiae—p. 138

Sponges are filter-feeding animals that grow in upright or encrusting forms; they are sessile as adults (though a few can glide about a little). In contrast to other multicellular animals, they lack organs or tissues and often are considered just a conglomeration of cells. Sponge construction builds on a framework of tiny skeletal spicules of calcium carbonate or silica/glass and structural protein fibers (collagen). Innumerable microscopic pores in the outer sponge surface carry water via tubes to tiny chambers lined with flagellum-whirling choanocytes ("collar cells") that propel water currents from which they efficiently filter out bacteria and other fine particles or absorb dissolved nutrients. From these chambers water flows through converging passages to one or a few large excurrent openings and so back into the sea. A fist-sized sponge may pump a liter of water per hour, extracting over 95 percent of the bacteria from that flow.

There are only a few cell types in sponges, and many can transform to other cell types. For example, in some sponges choanocytes can give rise to sperm (Anakina and Korotkova 1989). And sponges readily rearrange their cells and internal structures in response to hydrodynamic forces.

Most sponges are sequential hermaphrodites, functioning first as one sex, then the other. Sperm are shed into the water and eggs are usually retained. Most sponges also reproduce asexually, for example, by fragmentation or budding.

We provide an example from each of two sponge classes: calcareous sponges, which are usually small and have sharp, needlelike, simple or branched calcareous spicules; and demosponges, which have glassy spicules that are varied in shape and fibers of spongin, a structural protein. Almost 80 percent of known sponges are demosponges, including bath sponges, which, however, lack spicules.

PHYLUM CNIDARIA (Gk. *knidē*, nettle)

For Example:

Hydroids and hydromedusae—Class Hydrozoa—pp. 96, 210
Anemones, corals, zoanthids, sea pens, gorgonians—Class Anthozoa—pp. 218, 152, 202, 219, 198
Jellyfish—Class Scyphozoa, p. 117

The Cnidaria are a diverse group of predaceous carnivores. They all have unique cells, called cnidocytes, that contain structures (cnidae) of three types: nematocysts, spirocysts, and ptychocysts. Nematocysts are complex capsules containing coiled, hollow threads that typically are venomous and armed with barbs or spines. They are used for food capture or defense. Spirocysts function as sticky entanglers, and ptychocysts are used to build tubes.

Cnidarians are simple animals, radially symmetrical around a simple sac that serves both circulatory and digestive functions. The single opening acts as both mouth and anus. Cnidarians have two basic forms: a sessile, flowerlike polyp with nematocyst-armed tentacles, and a swimming umbrella-like medusa with tentacles hanging in a fringe from the margin, also nematocyst-laden. Cnidarians are constructed of two cell layers: ectoderm (= epidermis) and entoderm (= gastrodermis), separated by fibrous and gelatinous material that may be with or without cells. Thus they lack the mesoderm of more evolved animals. The cnidarian's simple nervous system is a diffuse network of nerve cells that are largely unpolarized—that is, nerve impulses can proceed from cell to cell in either direction—unlike the polarized cells typical of most other animals' nervous systems. It also lacks a brain. Nevertheless, this rudimentary system is enough to mediate such complex behavior as jellyfish swimming and feeding. Typically cnidarians have tiny motile, wormlike, ciliated planula larvae. The distinctive features of the three classes can be summarized as follows:

Hydrozoans are mostly small, delicate animals, often colonial, and often with two very different phases in their life cycles: an asexual hydroid phase of polyps that alternates with a sexual, very small hydromedusa phase (see fig. 10). Hydromedusae may be released or retained within the colony as greatly reduced medusoids. They in turn spawn eggs or sperm, and fertilization leads to planula larvae, which develop into another hydroid. The simple hydrozoan body cavity is undivided.

Anthozoans can be solitary or colonial. They exist only in the polyp phase, lacking a medusa or medusoid phase altogether. Their body cavity is divided by radiating septa laden with nematocysts. In forms without a calcareous skeleton, the cavity is distended with water and so is under slight pressure, which acts as a hydrostatic skeleton. Most anthozoa have separate sexes, and most reproduce both sexually and asexually. This immense and diverse class includes anemones, solitary corals and the various colonial soft and stony corals, zoanthids, sea pens, and gorgonians.

Scyphozoans are the animals we know as "jellyfish"—or, more accurately, medusae. Gametes appear and ripen in the gut wall. Often the medusae alternate with a very small, asexual polyp phase. Polyps split off tiny medusae by strobilation. Sense organs occur in notches at the periphery of the medusa's bell, alternating with nematocyst-containing tentacles.

PHYLUM MOLLUSCA (L. *molluscus*, soft)

For Example:

Snails—Class Gastropoda—p. 211
Clams and mussels—Class Bivalvia—pp. 239, 27
Chitons—Class Polyplacophora—p. 166
Squids and octopuses—Class Cephalopoda—p. 222

The evolution of molluscan shells is testimony to the relentless pressure exerted by predators on these soft-bodied animals. Molluscs are basically unsegmented (even if the shell is divided into repeated plates or some organs are serialized) and bilaterally symmetrical animals. Despite the big-body molluscan design, the coelom (a body cavity outside of the gut that houses internal organs) is generally tiny. Blood circulates mostly through porous lymphatic spaces, not extensive closed vessels, except in the highly active cephalopods. The "typical" molluscan body, though radically modified in several classes, is rather snail-like. It has a large dorsal visceral mass; an umbrella-like dorsal mantle spreading like a skirt over a broad, creeping foot; and a shell secreted by the epidermis of the mantle. The mouth (except in bivalves) usually has a muscular, scraping tongue, or radula.

All the diverse molluscan forms are variations on a basic scheme; scarcely a trait has escaped thorough remolding in one or another of the phylum's seven living classes. Molluscs seem to have radiated into virtually every available sort of habitat and way of life, though few are internal parasites. And different classes have converged in their evolutionary exploits. Thus, even snails range from open-ocean swimmers to completely sessile species. Some cephalopods rival fish in their athleticism and even their apparent intelligence. This phylum shows what extraordinary diversity can arise from a sufficiently plastic design, however limited and unpromising it would seem without the phylum's evolutionary accomplishments to prove otherwise.

Major characteristics of the four molluscan classes are as follows:

The gastropods—snails and sea slugs (nudibranchs)—develop a visceral hump twist of 180 degrees vis-à-vis the foot during larval development; this process, which is known as torsion, results in a twisted, asymmetric adult anatomy with the anus at the head end (see Fig. 14). Subsequent "detorsion" occurs in nudibranchs. Gastropods have a creeping foot and a caplike, often coiled dorsal shell, which in some lineages, such as slugs and nudibranchs, is lost.

Bivalve bodies are laterally compressed, and the foot has become a narrow digging tool. The shell is hinged into left and right valves. An evolutionary shift to filter-feeding exploits the elaborate gills that extend the length of the body on each side; those bivalves buried beneath the surface have inhalant and exhalant siphons. The head is reduced and the radula lost.

The chiton's body is flat and carries an eight-valved dorsal shell. The head is greatly reduced, with no eyes or tentacles. Repeated gills lie in grooves on both sides of the large foot. Chitons glide about very slowly; most species are herbivorous.

Cephalopods such as the squid and the octopus have an elaborate head and camera-like image-forming eyes. There is a well-developed nervous system and resulting complex behavior. Cephalopods are highly motile and predatory. Their foot is rolled into the muscular siphon used in jet propulsion. Grasping arms surround the beaked mouth. The shell usually is reduced or absent. Blood moves through an extensive closed circulatory system.

There are several other less famous molluscan classes, which we omit here because they are not represented in this book.

PHYLUM ANNELIDA (L. *annulus*, ring)

For Example :

Marine burrowers, crawlers, swimmers, and tube dwellers—Class Polychaeta—pp. 83, 105, 154

Annelids—which also include the earthworms (Class Oligochaeta), on land and in fresh water, and the often parasitic leeches (Hirudinida), in pools and streams—are bilaterally symmetrical segmented worms. Like the more evolved animals, annelids are constructed from three layers: the ectoderm, mesoderm, and entoderm. The worm's body is covered with a thin, flexible, relatively impermeable cuticle. The main unit of construction is the segment itself, repeated along the body's length, each separated from adjacent segments by septa. A ganglion of the nerve cord, organs such as kidneys, and gonads are repeated in each regular segment. The worm's gut, central nervous system, longitudinal muscle, and main blood vessels perforate the septa and run the length of the body. The first and last segments of annelids are modified into a head and "tail." Various specialized appendages protrude from specific segments.

The largest of the three classes of annelids is the strictly marine Polychaeta. Body cavities (coeloms) in these worms are simply paired lateral spaces within each segment; they provide hydrostatic support. The polychaetes' external anatomy and appearance generally reflect habits and habitat. Thus, "errant" (wandering) polychaetes have prominent lateral bristles projecting from stumpy "parapodia" (side feet) on each segment, photoreceptors as simple pigment cups, and, if they are predaceous carnivores, daunting jaws. In contrast, burrowers have reduced parapodia, peristaltic movements that aid digging, and meager sense organs. And tube-dwelling filter feeders have elaborate crowns of tentacles on the head or anterior trunk segments, some even with compound eyes. Most polychaetes release eggs and sperm into the sea, and a characteristic larval stage (the trochophore) follows fertilization.

PHYLUM ARTHROPODA (Gk. *arthro,* joint; L. *pod,* foot)

For Example:

Insects—Class Insecta—p. 18
Sea spiders—Class Pycnogonida—p. 53

Crabs, shrimps, copepods, barnacles, isopods, amphipods—Class Crustacea—
in various subclasses and orders—pp. 190, 126, 127, 162, 41, 106

The arthropods constitute a vast phylum with perhaps a million described species, four-fifths of them insects, mostly beetles. And undoubtedly there are millions more insects yet to be discovered. Members of this pervasive phylum, which share many features with the annelids, have invaded every habitable environment. Although many aspects of their phylogeny and evolution remain controversial, a discussion of such issues would not help us enjoy the creatures before us.

Their bodies consist of repeated segments each of which typically bears a pair of legs. The body—including the muscles (the part in crabs that tastes so good!)—is encased in a jointed exoskeleton secreted by the epidermis. In order for an arthropod to grow, its exoskeleton splits open and the animal wriggles out, a process called molting. The new, initially soft exoskeleton expands before it hardens.

Segmentation is clear enough in early development, but it is masked in adults by the way segments fuse into body regions—head, thorax, abdomen. Arthropods have marked cephalization with antennae, often well-developed compound eyes, and a brain consisting of two or three paired ganglia that extend to a ventral nerve cord. Arthropods often exhibit intricate behavior, exploiting their relatively complex nervous system and elaborate sense organs. Although the exoskeleton presents a barrier for sensory nerve endings in the epidermis, arthropods have various sensory hairs, bristles, pores, and slits that function in chemoreception and detect vibrations. The exoskeleton also poses a problem for respiration; many of the largely marine crustaceans get around this by having blood-filled gills, sometimes protected beneath a shield of the exoskeleton.

The taxonomy of this large group is complex, with several subphyla, some forty classes, and numerous subclasses, superorders, and orders. For our purposes it will suffice to list examples from various taxa that are represented in this book, together with just a word about principal identifying features.

We cover only two noncrustacean classes of arthropod: insects and sea spiders. Rove beetles—our insect representatives—have a tough exoskeleton and biting mouth parts; the two species herein lack wings. The gangly sea spiders, or pycnogonids, are mostly legs—four pairs of walking legs. The tiny body is not obviously differentiated into head, thorax, and abdomen.

The following taxa all are crustaceans, a diverse group named for their hard but flexible crust. Crustaceans have a shield, or carapace, over the head; their appendages have multiple joints.

Barnacles as adults are the only sessile arthropods. They can best be recognized as crustaceans during their larval stages.

Copepods are tiny swimmers whose long antennae and streamlined bodies give them a T-shape. Some species are among the most numerous animals on earth.

Amphipods are laterally compressed, while isopods are flattened. Neither has a carapace, and in both groups the first thoracic segment is fused with the head. They each have one pair of mouth parts and seven pairs of thoracic legs—legs quite similar to each other for the isopods, modified for different functions in amphipods. They have nonstalked compound eyes.

Euphausids, or krill, are shrimplike in appearance. They are extremely gregarious, and all are planktonic.

The decapods have a prominent hoodlike carapace that is fused to all the thoracic segments. So slight a trait, so huge a group! These are the crustaceans we all know best—crabs, lobsters, and "true" shrimp. All have stalked compound eyes, three highly modified thoracic appendages that serve as accessory feeding structures, as well as five pairs of thoracic legs—hence their name.

THE LOPHOPHORATES (Gk. *lophos*, crest; *pherein*, to carry)

For Example:

Phoronids—Phylum Phoronida—p. 65
Moss animals—Phylum Bryozoa—p. 84
Lamp shells—Phylum Brachiopoda—p. 233

These closely related phyla share many anatomical features, such that lumping them together in a superphylum the Lophophorates may be warranted. They all feed and respire with a specialized crown of ciliated tentacles surrounding the mouth, the lophophore, a feature that may have arisen independently in the evolution of each group. They live a benthic life, the wormlike phoronids in tubes, the bryozoans in compartmental exoskeletons, and the brachiopods in bivalved shells growing on stalks. In phoronids and bryozoans the anus is close to the mouth. Each has a U-shaped gut, an advantage for these enclosed animals that prevents them from "fouling their nest": waste is carried away by the lophophore's rejection current. Hinged brachiopods lack an anus; they expel waste from their mouth and eject it by clapping their dorsal and ventral shells. Bryozoans, superficially similar to bushlike or encrusting hydroids, are colonial and grow by budding. These animals all have coeloms, excretory organs, discrete gonads, and the other paraphernalia of "advanced" invertebrates.

Traditionally, anatomical traits have been used to align the lophophorates with deuterostome phyla such as echinoderms and chordates. More recent molecular studies suggest that lophophorates are more closely related to molluscs and annelids—protostome phyla (Halanych et al. 1995). The echinoderm/chordate and the arthropod/mollusc/annelid groups occupy different limbs on one of the most fundamental splits in the evolutionary tree. According to one study based on the "molecular clock," the divergence between the two groups (which are named for the way in which the embryonic mouth forms) may have occurred 670 million years ago, in late Precambrian times (Ayala, Rzhetsky, and Ayala 1998). It will take time to amass enough information to clarify these fascinating relationships.

PHYLUM ECHINODERMATA (Gk. *echinos,* spine; L. *derma,* skin)

For Example:

Seastars—Class Asteroidea—p. 29
Sea urchins—Class Echinoidea—p. 178
Sea cucumbers—Class Holothuroidea—p. 160
Brittle stars—Class Ophiuroidea—p. 188
Feather stars—Class Crinoidea—p. 236

Echinoderms today are pentamerously symmetrical as adults, but they are bilaterally symmetrical as larval young. The adult anatomy emerges during the metamorphosis that separates the larval and adult phases of development. Fossil forms were often bilaterally symmetrical or even helical as adults.

Tubes and chambers in the body wall constitute an elaborate, coelomic "water-vascular" system that hydraulically operates the hundreds of tube feet used in feeding, respiration, and locomotion. The body wall is stiffened by an endoskeleton of often tightly conjoined crystalline plates (ossicles). The external surface may carry numerous spines and pincers.

Despite their marked structural differences, echinoderms are closely allied with the chordates, as evidenced by the similar way the early embryonic mouth forms in the two phyla (deuterostome development)—a relationship recently confirmed by an extensive comparison of DNA sequences. Why, then, do chordates and echinoderms look so different from each other? It turns out that the same regulatory genes are put to different uses in the two phyla. For example, a gene (*Distal-less*) that is active in limb formation in mammals has been put to work forming tube feet in echinoderms (Panganiban et al. 1997).

Some characteristics of the five most familiar echinoderm classes are as follows:

The seastars are mostly sluggish predators, using stout arms and often strongly adhering tube feet to subdue and steadily dismantle prey. To feed they often evert their stomach outside of the body and apply it to their prey.

Sea urchins are generally stiff and globular, covered with spines and tube feet. Their jaw apparatus is exceedingly complex despite their merely algal-feeding, detritus-scavenging, and substrate-scraping habits.

Sea cucumbers are like stout worms but with telltale tracts of tube feet and widely separated ossicles that may have intricate, anchorlike shapes. Their anterior tube feet enlarge into ten elaborate sticky tentacles that are used in suspension feeding or gathering detritus.

The body of brittle stars usually comprises a small central disk within which the gut, coelom, and gonads are confined. Short tube feet protrude from the long, slender, bony arms. They move by rowing with their arms, not by pacing with tube feet. Brittle stars may cast off arms when they are disturbed. They are suspension feeders, detritivores, and scavengers.

Feather stars have body proportions roughly that of brittle stars—a condensed central body surrounded by many long, feathery legs that capture suspended particles for food. The body's posture, however, is inverted: the mouth faces up. In sessile sea lilies, a dorsal stalk attaches fast to the substrate; in motile forms, several flexible dorsal cirri grip the bottom like fingers.

PHYLUM CHORDATA (Gk. *chordē,* string)

For Example:

Sea squirts, salps—Subphylum Tunicata—pp. 164, 121
Fish, birds, mammals—Subphylum Vertebrata—pp. 80, 21, 78

All chordates, at least in some stage of their development, have the same traits: the notochord for which the phylum is named (a flexible, turgid rod that runs the length of the back); pharyngeal gill slits; a dorsal tubular nerve cord; an endostyle (glandular groove in the pharynx) in the tunicates or a thyroid gland in vertebrates; and a post-anal muscular tail. Finally, during early embryonic development the mouth forms secondarily (deuterostome development). The coelom is lost in the tunicates.

Ascidians, such as the filter-feeding "sea squirts," exemplify the chordate traits only during their larval stage, when they resemble tiny tadpoles. With metamorphosis into sessile adults, ascidians resorb most of the nervous system, the notochord, and the tail. Salps, pelagic tunicates that lack tadpole larvae, never show these latter chordate traits at all. The pharyngeal basket used for filter feeding in most tunicates is elaborated from the larval gill slits.

Tunicates and vertebrates could scarcely differ more yet still be in the same phylum! Tunicates lack a coelom, lack hard parts and limbs, lack eyes or other sophisticated sense organs, and almost all (one strange exception lives in Monterey Canyon) gain their food by using their pharyngeal gills to filter-feed minute plankton from the water. They have odd organs such as a reversing heart that pumps blood now one way, now the other. In some ascidians this blood is richer in vanadium than are commercial ores! The tunic of many species contains cellulose, a carbohydrate otherwise found only in seaweeds and plants.

As we just mentioned, vertebrate embryos, as well as tunicate tadpoles, exemplify the characteristic chordate traits. Every biology student has seen pictures of a human embryo, a pig embryo, a shark embryo: almost identical little comma-shaped bodies with a large head, prominent gill structures, and a tail. A notochord and tubular nerve cord lie beneath their bulging back. At this stage, the human embryo's circulation even resembles that of the shark, with a tubular heart that feeds forward into gill arches. Darwin, Haeckel, and others advanced the then radical idea that the striking similarities of these embryos suggest the evolution of these organisms from a common ancestry.

During vertebrate development, the iodine-rich thyroid gland develops from an embryonic gill structure. The cephalic end of the neural tube dilates into the relatively huge brain lobes and later becomes encased in a skull. The notochord and nerve cord become encased in a segmented spinal column, and the notochord is lost. In the various vertebrate classes, specializations occur adapting these animals to their respective lifestyles—in fish, for example, gills develop from embryonic gill slits, and so forth. Vertebrates are often intensely visual animals, besides exploiting sensitivity to noise, smell and taste, and pressure. Most of them are active foragers. They are behaviorally intricate animals, matched (after a fashion) only by some cephalopod molluscs.

And so this review of several phyla brings us at last to ourselves, enchanted observers of this living bay, and to the vertebrate animals out there whose camp and corner we share.

GLOSSARY

The glossary does not include names of phyla, classes, or subphyla. See the appendix for a discussion of the phyla covered in this book.

aboral	The side of the animal opposite the mouth.
abyssal	Depths of the ocean below the continental shelf.
acontia	In cnidarians, long threadlike processes extruded from a partition of the body wall and loaded with nematocysts.
alternation of generations	In a life cycle, the alternation of a phase that reproduces sexually with one that reproduces asexually.
ancestrula	The founding (first) zooid of a bryozoan colony.
angiosperm	A flowering plant; the angiosperms are a major division of the plant kingdom.
ascidian	A sea squirt, member of the tunicate subphylum. Virtually all are sessile in the adult stage.
baleen	Horny plates (whalebone) of keratin that grow from the upper jaw in baleen whales; used to strain water and capture plankton.
Batesian mimicry	The resemblance of one animal (the mimic) to another (the model) to the benefit of the mimic when the model is dangerous. Set forth by the English naturalist H. W. Bates while working with butterflies.
benthic	Pertaining to the sea bottom.
budding	Reproduction by the development of an outgrowth or bud.
canthaxanthin	In isopods, a red molecule obtained by adding oxygen to beta carotene.
carapace	A single plate covering the fused head and thorax of some arthropods.

carotenoid	Red, yellow, and orange plant pigments chemically and functionally similar to carotene. Some are active photosynthetically.
carposporophyte	One of three generations in a red alga.
cephalization	Increasing importance of the front (head) end in animal development; concentration of neural and sensory functions toward the head.
ceras (pl. cerata)	Dorsal processes on eolid nudibranchs, outgrowths of the gut that function as gills.
chemoautotroph	An organism deriving energy from the oxidation of inorganic compounds (rather than from the sun's energy) and using carbon dioxide as the principal carbon source.
chemosynthesis	The synthesis of organic compounds with energy derived from inorganic chemical reactions.
chitin	A structural carbohydrate component of arthropod exoskeletons.
chlorophyll	A green photosynthetic pigment in plants and algae.
chloroplast	An inclusion in plant cells that contains chlorophyll and sometimes other pigments.
choanocyte	A cell with a collar around the base of a whiplike hair, as in sponges; also called "collar cell."
chromatophore	A pigment cell or group of cells, under control of the sympathetic nervous system, that can be altered in shape to produce a color change.
cilia (sing. cilium)	Small hairlike processes on cell surface.
cirrus (pl. cirri)	Name given to various tentacle-like appendages in different animals.
cloaca	In certain organisms, the common opening of the intestinal, genital, and urinary canals.
clone	A group of genetically identical organisms derived from a single individual by replication (i.e., asexual reproduction).
cnida (pl. cnidae)	In cnidarians, a unique intracellular capsule: a nematocyst, a ptychocyst, or a spirocyst.
coelenteron	The body cavity or gastrovascular cavity in cnidarians (coelenterates); it carries out digestive and circulatory functions.

coelom	A body cavity between the body wall and the gut, within the mesoderm; lacking in sponges, cnidarians, comb jellies, and flatworms.
collagen	A structural protein in connective tissue.
colony	A group of organisms with organic connections between them (e.g., bryozoans and some tunicates).
commensal	See *symbiosis.*
conceptacle	A fertile cavity in an alga that bears gametes.
ctenophore	One of a phylum of marine organisms with eight ciliary comb rows, sometimes called comb jellies.
cuticle	The hard external covering of many invertebrates, usually impervious to water.
deuterostome	A bilateral animal in which, developmentally or phylogenetically, the mouth forms secondarily, not from the initial opening in an early embryo. One of two major divisions of the animal kingdom (cf. *protostome*). Echinoderms and chordates are deuterostomes.
diatom	A microscopic alga with a silicon-dioxide skeleton.
dinoflagellate	A microscopic unicellular planktonic alga with two flagella.
dioecious	Having separate sexes.
diploid	Having two sets of chromosomes.
direct development	Development without a larval stage.
ectoderm	The outermost of the three primary embryonic layers and the source of neural tissue and sense organs.
embryo	An early stage in an organism's development.
endoderm	The innermost of the three primary embryonic layers and the source of the gut.
endoskeleton	An internal skeleton.
endostyle	A glandular groove in the pharynx of tunicates, the precursor of the vertebrate thyroid gland.
epiphyte	An alga or plant that lives on other algae or plants but does not derive water or nourishment from them.
epizoic	A plant or animal living on the surface of another animal.
eukaryote	An organism whose DNA is organized within the cells' nuclei.

excurrent	Flowing out.
exoskeleton	An external skeleton; e.g., the shell of a crab.
flagellum (pl. flagella)	A long whiplike fiber that rotates, propelling an animal.
fucoxanthin	An accessory photosynthetic pigment in giant kelp, rockweeds, and diatoms, for example. It gives giant kelp its golden brown color.
fusibility	The ability to recognize a cell's surface protein as "self" and therefore join with it.
gametophyte	The haploid, gamete-producing phase in an alga's or plant's life history.
ganglion	A mass of nerve cell bodies giving rise to nerve fibers.
genotype	The genetic constitution of an individual.
glycoprotein	A protein combined with a carbohydrate.
gnathopod	The arthropod limb near the oral cavity that is modified to handle food.
gonad	Primary sex gland: ovary or testis.
gonozooid	A zooid containing a gonad.
haploid	Having a single set of chromosomes per nucleus.
hermaphrodite	Having both male and female sex organs. *Sequential hermaphrodite:* in which sex organs function one following the other; *simultaneous hermaphrodite:* in which both sex organs function at the same time.
heteromorphic	Having different forms at different times.
hydromedusa	The medusa (jellyfish) phase in the life history of a hydrozoan.
hydrostatic pressure	Water pressure. May serve as skeletal support in some invertebrates.
incurrent	Admitting water.
indirect development	Having a free-living larval stage in the early course of development.
interleukin	Any of several proteins released by certain cells that act as agents of communication between different cells, such as promoting inflammation or enhancing cell proliferation.
isomorphic	Superficially alike, similar in appearance.
keratin	A horny material (scleroprotein) of dead cells forming the basis of epidermal structures such as fingernails and whale baleen.

larva	An embryo that becomes self-sustaining and independent before assuming its parents' features.
lecithotrophic	Feeding on stored yolk.
lipoprotein	A protein combined with lipids.
lophophore	The specialized filter feeding organ bearing tentacles found in bryozoans, phoronids, and brachiopods.
mantle	A fleshy covering of molluscs and brachiopods that lines the shell, enclosing the internal organs.
mantle cavity	In a mollusc, the space between the mantle lining the shell and the body, where the respiratory organs lie.
medusa	The free-swimming sexual "jellyfish" body form of cnidarians.
medusoid	In hydrozoan colonies, the retained medusa (jellyfish) phase of development.
meiofauna	Benthic organisms less than 0.5 mm in size; often called "interstitial fauna."
meiosis	A form of cell division in which the chromosome number is reduced from diploid to haploid; reduction division.
mesentery	A muscular partition extending inward from the body wall in cnidarians; a peritoneal fold serving to hold viscera in position.
mesoderm	The middle of the three primary embryonic layers, giving rise to skeletal and muscular tissues.
metamorphosis	A more or less abrupt change of form and structure undergone by an animal during its development.
micron	One-thousandth of a millimeter (10^{-6} m).
mitosis	Cell division in which each daughter nucleus has the same number of chromosomes as the parent nucleus.
monophyletic	Derived from a single common ancestral form.
nanometer	One-millionth of a millimeter (10^{-9} m).
nanoplankton	Plankton with a size range from 2 to 20 microns.
nekton	Actively swimming open-water animals.
nematocyst	A cnida in cnidarian animals consisting of a hollow thread armed with venomous spines or barbs that can shoot out following stimulation.
notochord	A firm supporting rod in the back of the animal between the nerve cord and gut, present in all chordates at some time

in development. It is replaced by the vertebral column in vertebrates.

ommatidium (pl. ommatidia)	An element of a compound eye, consisting of a corneal lens, crystalline cone, refractive rod, pigment cells, and sheath.
operculum	A horny lidlike structure attached to the foot of some gastropods that closes the shell's opening when the foot is withdrawn.
ossicle	Any small bone, e.g. a small calcareous plate embedded in the echinoderm body wall for strengthening.
oviparous	Egg laying.
ovipore	Opening of the oviduct.
ovoviviparous	Producing an egg that hatches within the maternal body.
oxygen minimum zone (OMZ)	The ocean depth between 500 and 1000 m where the oxygen content is greatly reduced, even approaching zero.
pedicellariae	Pincerlike structures on the backs of some echinoderms.
pelagic	Occurring in the open water of the world's oceans.
pentamerous	Five-part construction, found in seastars and their kin.
peptide	Simply, a fragment of protein consisting of a chain of amino acids.
pharyngeal basket	In tunicates, the pharyngeal structure with slits that filters water to obtain food and oxygen.
pheromone	A chemical secreted by a species that influences the behavior of others of the same species.
photosynthesis	In green plants and algae, the process whereby the energy of sunlight is used to synthesize organic compounds from carbon dioxide and water.
phycoerythrin	A photosynthetic pigment that gives the red color to red algae.
phytoplankton	Drifting algal plankton, including diatoms and dinoflagellates.
pigment cup	Simple depression with photoreceptor cells but lacking specialized structures such as lens and iris.
plankton	Animals and plants drifting in the ocean, usually small or microscopic.
planktotrophic	Feeding on plankton.
planula	The free-living larva of a cnidarian.
polymorphism	Different forms of organs or of individuals in the same species.

polyp	In cnidarians, an individual having a tubular body with a mouth-anus opening surrounded by a ring or rings of tentacles; a zooid or individual of a colonial animal.
proboscis	The extensible tubular process of the oral region of a gastropod, usually containing the radula.
protostome	A member of one of two major branches of the animal kingdom in which the mouth derives developmentally from the initial opening in the early embryo (cf. *deuterostome*). Molluscs, annelids, and arthropods are protostomes.
ptychocyst	A cnida in tube anemones that ejects the adhesive threads used to build the animals' tubes.
radula	A horny ribbon with rows of teeth used in rasping food, in most molluscs.
respiration (cellular)	Process by which energy is acquired in an organism by the breakdown of storage molecules (sugars) accompanied by the release of carbon dioxide, in the presence or absence of oxygen (i.e., aerobically vs. anaerobically).
rhinophores	The sensory tentacles on the head of many nudibranchs; thought to have an olfactory function.
rhodopsin	Visual pigment, a light-sensing molecule in all known photoreceptors.
septa	Dividing walls or membranes.
sessile	Attached or stationary, as opposed to free-living or motile.
siphonophore	A pelagic hydrozoan that has polymorphic colonies of both polyps and medusoid forms and bears floats.
sorus (pl. sori)	An area where spores are produced (e.g., on a kelp blade).
spermatophore	A capsule containing a number of sperm; usually formed within a specialized part of the male reproductive system.
spicule	A minute pointed process, silicious or calcareous, found in sponges.
spirocyst	A cnida in anthozoan cnidarians that when stimulated ejects a long, sticky, entangling thread.
sporophyte	The diploid, spore-producing phase in an alga's or plant's life history.
stolon	A tubular runner that connects zooids.

strobilate	In cnidarians and tapeworms, asexual reproduction by transverse division of the body so segments break free as independent organisms.
substratum	A base on which an organism lives.
swim bladder	A gas-filled bladder giving buoyancy control to most bony fish.
symbiosis	The living together of two species, including *commensalism,* in which one organism gains at no cost to the other, and *mutualism,* in which both organisms benefit.
taxon (pl. taxa)	Any definite unit in classification of plant and animals; a taxonomic unit.
test	Shell or hardened outer covering; the endoskeleton of certain echinoderms (e.g., urchins, sand dollars).
thallus	The body of an alga.
trochophore	Free-swimming larval stage of annelids and some molluscs in which the body is ringed by a girdle of cilia.
tunic	The tough gelatinous tissue that protects and supports the body of a tunicate.
veliger	The second stage in larval life of certain molluscs, developed from a trochophore.
viviparous	Producing living young rather than eggs.
zooid	A single individual in a colony of animals, particularly in bryozoans, cnidarians, and tunicates.
zooplankton	Animals free-floating or drifting in the open water.
zooxanthellae	Yellow or brown unicellular algae living in other animals.
zygote	Cell formed by union of two gametes or reproductive cells; fertilized egg.

REFERENCES

Abbott, Donald P. 1987. *Observing marine invertebrates: Drawings from the laboratory.* Ed. Galen H. Hilgard. Stanford: Stanford University Press.

Abbott, Donald P., and Eugene C. Haderlie. 1980. Prosobranchia: Marine snails. In *Intertidal invertebrates of California,* ed. Robert H. Morris, Donald P. Abbott, and Eugene C. Haderlie, 230–307. Stanford: Stanford University Press.

Abbott, Donald P., and Andrew Todd Newberry. 1980. Urochordata: The tunicates. In *Intertidal invertebrates of California,* ed. Robert H. Morris, Donald P. Abbott, and Eugene C. Haderlie, 177–226. Stanford: Stanford University Press.

Abbott, Donald P., and Donald J. Reish. 1980. Polychaeta: The marine annelid worms. In *Intertidal invertebrates of California,* ed. Robert H. Morris, Donald P. Abbott, and Eugene C. Haderlie, 448–489. Stanford: Stanford University Press.

Abbott, I. A., and G. J. Hollenberg. 1976. *Marine algae of California.* Stanford: Stanford University Press.

Ajeska, Richard A., and James Nybakken. 1976. Contributions to the biology of *Melibe leonina* (Gould 1852). *Veliger* 19 (1): 19–26.

Alexander, R. McNeill. 1999. One price to run, swim, or fly? *Nature* 397: 651–653.

Alkon, D. L. 1993. GABA-mediated synaptic interaction between the visual and vestibular pathways of *Hermissenda. J. Neurochem.* 61 (2): 556–566.

Alldredge, A. L., and L. P. Madin. 1982. Pelagic tunicates: Unique herbivores in the marine plankton. *BioScience* 32 (8): 655–663.

Amsler, C. D., and M. Neushul. 1989. Diel periodicity of spore release from the kelp *Nereocystis luetkeana* (Mertens) Postels et Ruprecht. *J. Exp. Mar. Bio. Ecol.* 134 (2): 117–128.

Anakina, R. P., and G. P. Korotkova. 1989. Spermatogenesis in *Leucosolenia complicata* Mont., Barents sea sponge. *Ontogenez* 20 (1): 77–86. [In Russian]

Anderson, M. E. 1994. Systematics and osteology of the Zoarcidae (Teleostei: Perciformes). *Ichthyol. Bull. J.L.B. Smith Inst. Ichthyol.* 60: 1–120.

Anderson, P. A. V., and Q. Bone. 1980. Communication between individuals in salp chains. II. Physiology. *Proc. R. Soc. Lond. B* 210: 559–574.

Annett, Cynthia, and Raymond Pierotti. 1984. Foraging behavior and prey selection of the leather seastar *Dermasterias imbricata. Mar. Ecol. Prog. Ser.* 14: 197–206.

Aoki, Masakazu. 1997. Comparative study of mother-young association in caprellid amphipods: Is maternal care effective? *J. Crustac. Biol.* 17 (3): 447–458.

Avila, Conxita. 1998. Chemotaxis in the nudibranch *Hermissenda crassicornis:* Does ingestive conditioning influence its behaviour in a Y-maze? *J. Moll. Stud.* 64: 215–222.

Ayala, Francisco José, Andrey Rzhetsky, and Francisco J. Ayala. 1998. Origin of the metazoan phyla: Molecular clocks confirm paleontological estimates. *Proc. Natl. Acad. Sci. USA* 95: 606–611.

Bahamondes-Rojas, I., and M. Dherbomez. 1990. Partial purification of glycoconjugates capable of inducing metamorphosis in competent larvae of the nudibranch mollusk *Eubranchus doriae* (Trinchese, 1879). *J. Exp. Mar. Biol. Ecol.* 144 (1): 17–27. [In French]

Baines, George Whitney. 1979. Blood pH effects in eight fishes from the teleostean family Scorpaenidae. In *Readings in ichthyology,* ed. Milton S. Love and Gregor M. Cailliet, 190–199. Santa Monica, CA: Goodyear.

Bakus, Gerald J., and Donald P. Abbott. 1980. Porifera: The sponges. In *Intertidal invertebrates of California,* ed. Robert H. Morris, Donald P. Abbott, and Eugene C. Haderlie, 21–39. Stanford: Stanford University Press.

Banaszak, A. T., R. Iglesias-Prieto, and R. K. Trench. 1993. *Scrippsiella velellae,* sp. nov. (Peridiniales), and *Gloeodinium viscum,* sp. nov. (Phytodiniales): Dinoflagellate symbionts of two hydrozoans (Cnidaria). *J. Phycol.* 29 (4): 517–528.

Barbieri, Elena, Kevin Barry, Alice Child, and Norman Wainwright. 1997. Antimicrobial activity in the microbial community of the accessory nidamental gland and egg cases of *Loligo pealei* (Cephalopoda: Loliginidae). *Biol. Bull.* 193: 275–276.

Barnes, Anthony T., and James F. Case. 1972. Bioluminescence in the mesopelagic copepod *Gaussia princeps* (T. Scott). *J. Exp. Mar. Biol. Ecol.* 8: 53–71.

Barnes, J. R., and J. J. Gonor. 1973. The larval settling response of the lined chiton *Tonicella lineata. Mar. Biol.* 20: 259–264.

Barnes, R. D. 1980. *Invertebrate zoology*. 4th ed. Philadelphia: Saunders College.

Barry, J. P., C. H. Baxter, R. D. Sagarin, and S. E. Gilman. 1995. Climate-related, long-term faunal changes in a California rocky intertidal community. *Science* 267: 672–675.

Barry, J. P., H. G. Greene, D. L. Orange, C. H. Baxter, B. H. Robison, R. E. Kochevar, J. W. Nybakken, D. L. Reed, and C. M. McHugh. 1996. Biologic and geological characteristics of cold seeps in Monterey Bay, California. Deep-sea research, part I. *Oceanographic Research Papers* 43 (11–12): 1739–1762.

Basch, L. V., and J. S. Pearse. 1996. Consequences of larval feeding environment for settlement and metamorphosis of a temperate echinoderm. *Oceanologica Acta* 19 (3–4): 273–285.

Bauer, R. T. 1977. Antifouling adaptations of marine shrimp (Crustacea: Decapoda; Caridea): Functional and adaptive significance of antennular preening by the third maxillipeds. *Mar. Biol.* 40: 261–276.

Beck, Gregory, and Gail S. Habicht. 1996. Immunity and the invertebrates: The evolution of the immune system. *Sci. Am.,* Nov.: 60–66.

Beebe, William. 1934. *Half Mile Down.* New York: Harcourt, Brace & Co.

Bennett, B. A., C. R. Smith, B. Glaser, and H. L. Maybaum. 1994. Faunal community structure of a chemoautotrophic assemblage on whale bones in the deep northeast Pacific Ocean. *Mar. Ecol. Progr. Ser.* 108 (3): 205–223.

Bergeron, Lou. 1997. Geology. In *Natural history of the Monterey Bay National Marine Sanctuary,* 28–55. Monterey, CA: Monterey Bay Aquarium.

Bernstein, Brock B., and Nancy Jung. 1979. Selective pressures and coevolution in a kelp canopy community in southern California. *Ecol. Monogr.* 49 (3): 335–355.

Bickell, L. R., and F. S. Chia. 1979. Organogenesis and histogenesis in the planktotrophic veliger of *Doridella steinbergae* (Opisthobranchia: Nudibranchia). *Mar. Biol.* 52: 291–313.

Bickell, L. R., F. S. Chia, and B. J. Crawford. 1981. Morphogenesis of the digestive system during metamorphosis of the nudibranch *Doridella steinbergae* (Gastropoda): Conversion from phytoplanktivore to carnivore. *Mar. Biol.* 62: 1–16.

Bickell-Page, L. R. 1991. Repugnatorial glands with associated striated muscle and sensory cells in *Melibe leonina* (Mollusca, Nudibranchia). *Zoomorphology (Berl.)* 110 (5): 281–292.

Bieri, Robert. 1977. The ecological significance of seasonal occurrence and growth rate of *Velella* (Hydrozoa). *Publ. Seto Mar. Biol. Lab.* 24 (1/3): 63–76.

Bone, Q., P. A. V. Anderson, and A. Pulsford. 1980. The communication between individuals in salp chains. I. Morphology of the system. *Proc. R. Soc. Lond. B* 210: 549–558.

Borowsky, Betty. 1985. Differences in reproductive behavior between two male morphs of the amphipod crustacean *Jassa falcata* Montagu. *Physiol. Zool.* 58 (5): 497–502.

Bouget, F.-Y., F. Berger, and C. Brownlee. 1998. Position-dependent control of cell fate in the *Fucus* embryo: Role of intercellular communication. *Development* (Cambridge) 125 (11): 1999–2008.

Brandt, Johanne Friderico. 1835. *Prodromus descriptionis animalium ab H. Mertensio in orbis terrarum circumnavigatione observatorum.* Fascic. 1, p. 13. Petropoli: Sumptibus Academiae [St. Petersburg: Acad. imp. sci.].

Bray, Richard N., and Alfred W. Ebeling. 1975. Food, activity, and habitat of three "picker-type" microcarnivorous fishes in the kelp forests off Santa Barbara, California. *Fish. Bull. (Wash. D.C.)* 73 (4): 815–829.

Bray, Richard N., and Mark A. Hixon. 1978. Night-shocker: Predatory behavior of the Pacific electric ray (*Torpedo californica*). *Science* 200: 333–334.

Brusca, R. C., and G. J. Brusca. 1990. *Invertebrates.* Sunderland, MA: Sinauer Assoc.

Burke, R. D., and R. F. Watkins. 1991. Stimulation of starfish coelomocytes by interleukin-1. *Biochem. Biophys. Res. Commun.* 180 (2): 579–584.

Butler, T. H. 1980. Shrimps of the Pacific coast of Canada. *Can. Bull. Fish. Aquat. Sci.*, no. 202: 47–49. Ottawa: Dept. of Fisheries and Oceans.

Caine, Edsel A. 1991. Reproductive behavior and sexual dimorphism of a caprellid amphipod. *J. Crustac. Biol.* 11 (1): 56–63.

Carefoot, Thomas. 1977. *Pacific seashores: A guide to intertidal communities.* Seattle: University of Washington Press.

Carter, John W., and David Behrens. 1980. Gastropod mimicry by another pleustid amphipod in central California. *Veliger* 22 (4): 376–377.

Cary, S. C., and S. J. Giovannoni. 1993. Transovarial inheritance of endosymbiotic bacteria in clams inhabiting deep-sea hydrothermal vents and cold seeps. *Proc. Natl. Acad. Sci. USA* 90 (12): 5695–5699.

Chadwick, Nanette E. 1987. Interspecific aggressive behavior of the corallimorpharian *Corynactis californica* (Cnidaria: Anthozoa): Effects of sympatric corals and sea anemones. *Biol. Bull.* 173: 110–125.

———. 1991. Spatial distribution and the effects of competition on some temperate Scleractinia and Corallimorpharia. *Mar. Ecol. Prog. Ser.* 70: 39–48.

Chadwick, N. E., and C. Adams. 1991. Locomotion, asexual reproduction, and killing of corals by the corallimorpharian *Corynactis californica. Hydrobiologia* 216/217: 263–269.

Chandrasekaran, Ravi. 1991. Endogenous rhythms in the soft coral *Anthomastus ritterii.* MS., Hopkins Marine Station, Stanford University, Pacific Grove, CA.

Chen, D. S., G. Van Dykhuizen, J. Hodge, and W. F. Gilly. 1996. Ontogeny of copepod predation in juvenile squid (*Loligo opalescens*). *Biol. Bull.* 190 (1): 69–81.

Childress, James J. 1975. The respiratory rates of midwater crustaceans as a function of depth of occurrence and relation to the oxygen minimum layer off southern California. *Comp. Biochem. Physiol.* 50A: 787–799.

Chung, J.-M. 1995. Dopamine as a strong candidate for a neurotransmitter in a hydrozoan jellyfish. *J. Biochem. Mol. Biol.* 28 (4): 323–330.

Clark, R. F., S. R. Williams, S. P. Nordt, and A. S. Manoguerra. 1999. A review of selected seafood poisonings. *Undersea Hyper. Med.* 26 (3): 175–184.

Coe, Wesley R. 1953. Influence of association, isolation, and nutrition on the sexuality of snails of the genus *Crepidula*. *J. Exp. Zool.* 122: 5–20.

Coelho, L., J. Prince, and T. G. Nolen. 1998. Processing of defensive pigment in *Aplysia californica:* Acquisition, modification, and mobilization of the red algal pigment r-phycoerythrin by the digestive gland. *J. Exp. Biol.* 201: 425–438.

Conlan, Kathleen E. 1989. Delayed reproduction and adult dimorphism in males of the amphipod genus *Jassa* (Corophioidea: Ischyroceridae): An explanation for systematic confusion. *J. Crustac. Biol.* 9 (4): 601–625.

Connor, Valerie M., and James F. Quinn. 1984. Stimulation of food species growth by limpet mucus. *Science* 225*:* 843–844.

Coombe, Deirdre R., Peter L. Ey, Charles R. Jenkins. 1984. Self/non-self recognition in invertebrates. *Q. Rev. Biol.* 59: 231–255.

Cornwall, Roger, Barbara Holland Toomey, Shannon Bard, Corrine Bacon, Walter M. Jarman, and David Epel. 1995. Characterization of multixenobiotic/multidrug transport in the gills of the mussel *Mytilus californianus*, and identification of environmental substrates. *Aquat. Toxicol.* 31: 277–296.

Cowles, D. L. 1994. Swimming dynamics of the mesopelagic vertically migrating penaeid shrimp *Sergestes similis:* Modes and speeds of swimming. *J. Crustac. Biol.* 14 (2): 247–257.

Cox, Paul Allen. 1993. Water-pollinated plants. *Sci. Am.,* Oct.: 68–74.

Crane, Jules M., Jr. 1969. Mimicry of the gastropod *Mitrella carinata* by the amphipod *Pleustes platypa. Veliger* 12 (2): 200.

Crisp, D. J. 1990. Gregariousness and systematic affinity in some North Carolinian [USA] barnacles. *Bull. Mar. Sci.* 47 (2): 516–525.

Croll, Don. 1998. Whales, krill, and canyons. Lecture, Monterey Bay Aquarium, July 7, 1998.

Cuoc, C., D. Defaye, M. Brunet, R. Notonier, and J. Mazza. 1997. Female genital structures of Metridinidae (Copepoda: Calanoida). *Mar. Biol.* 129: 651–665.

Darwin, Charles. 1839. *Journal of researches into the geology and natural history of the various countries visited by the H.M.S.* Beagle. London: Henry Colburn, Great Marlborough Street. Facsimile reprint, New York: Hafner, 1952.

———. 1854. *A monograph on the sub-class cirripedia.* London: Printed for the Ray Society.

Dawkins, Richard. 1989. *The selfish gene.* Oxford: Oxford University Press.

Dawsen, E. Yale. 1966. *Marine botany.* New York: Holt, Rinehart & Winston.

Dayton, Paul K. 1973. Two cases of resource partitioning in an intertidal community: Making the right prediction for the wrong reason. *Am. Nat.* 107: 662–670.

Degnan, B. M., S. M. Degnan, A. Giusti, and D. E. Morse. 1995. A hox hom homeobox gene in sponges. *Gene* (Amsterdam) 155 (2): 175–177.

Dickson, L. G., and J. R. Waaland. 1985. *Porphyra nereocystis:* A dual-daylength seaweed. *Planta* 165: 548–553.

Drew, K. M. 1949. Conchocelis phase in the life history of *Porphyra umbilicalis. Nature* 164: 748.

Duke, Richard C., David M. Ojcius, and John Ding-E Young. 1996. Cell suicide in health and disease. *Sci. Am.,* Dec.: 80–87.

Dunn, D. F. 1977. Dynamics of external brooding in the sea anemone *Epiactis prolifera. Mar. Biol.* 39: 41–49.

Edmands, S., and D. C. Potts. 1997. Population genetic structure in brooding sea anemones (*Epiactis* spp.) with contrasting reproductive modes. *Mar. Biol.* 127: 485–498.

Edwards, D. Craig. 1969. Predators on *Olivella biplicata*, including a species-specific predator avoidance response. *Veliger* 11 (4): 326–333.

Efford, Ian E. 1967. Neoteny in sand crabs of the genus *Emerita* (Anomura: Hippidae). *Crustaceana* 13: 81–93.

Epel, David. 1976. Sperm-egg interactions. *Oceanus* 19 (2): 34–38.

Evans, William G. 1980. Insecta, Chilopoda, and Arachnida: Insects and allies. In *Intertidal invertebrates of California,* ed. Robert H. Morris, Donald P. Abbott, and Eugene C. Haderlie, 641–658. Stanford: Stanford University Press.

Fadlallah, Jusef H. 1981. The reproductive biology of three species of corals from central California. Ph.D. diss., University of California, Santa Cruz.

Faulkes, Z., and D. H. Paul. 1997. Digging in sand crabs (Decapoda, Anomura, Hippaoidea): Interleg coordination. *J. Exp. Biol.* 200 (4): 793–805.

Faulkner, D. John, and Michael T. Ghiselin. 1983. Chemical defense and evolutionary ecology of dorid nudibranchs and some other opisthobranch gastropods. *Mar. Ecol. Prog. Ser.* 13: 295–301.

Faurot, Ellen R., Daniel P. Costa, and Jack A. Ames. 1986. Analysis of sea otter, *Enhydra lutris*, scats collected from a California haul-out site. *Mar. Mamm. Sci.* 2 (3): 223–227.

Feder, Howard M. 1955. On the methods used by the starfish *Pisaster ochraceus* (Brandt) in opening three types of bivalve molluscs. *Ecology* 36: 764–767.

Fenchel, T. 1988. Marine planktonic food chains. *Ann. Rev. Ecol. Sys.* 19: 1938.

Fenical, William, R. K. Okuda, M. M. Bandurraga, P. Culver, and R. S. Jacobs. 1981. Lophotoxin: A novel neuromuscular toxin from Pacific sea whips of the genus *Lophogorgia. Science* 212: 1512–1514.

Fernald, Russell D. 1997. The evolution of eyes. *Brain Behav. Evol.* 50 (4): 253–259.

Fields, W. G. 1965. The structure, development, food relations, reproduction, and life history of the squid *Loligo opalescens* Berry. *Calif. Dep. Fish Game Fish. Bull.* 131.

Fishlyn, Debby A., and David W. Phillips. 1980. Chemical camouflaging and behavioral defenses against a predatory seastar by three species of gastropods from the surfgrass *Phyllospadix* community. *Biol. Bull.* 158: 34–48.

Flegal, A. R., K. J. R. Rosman, and M. D. Stephenson. 1987. Isotope systematics of contaminant leads in Monterey Bay (California, USA). *Environ. Sci. Technol.* 21 (11): 1075–1079.

Foster, B. A. 1987. Barnacle ecology and adaptation. In *Barnacle biology,* ed. Alan J. Southward, 113–133. Crustacean Issues 5. Rotterdam: A. A. Balkema.

Fox, Denis L., and Donald W. Wilkie. 1970. Somatic and skeletally fixed carotenoids of the purple hydrocoral *Allopora californica. Comp. Biochem. Physiol.* 36: 49–60.

Francis, Lisbeth. 1976. Social organization within clones of the sea anemone *Anthopleura elegantissima. Biol. Bull.* 150: 361–376.

———. 1985. Design of a small cantilevered sheet: The sail of *Velella velella. Pac. Sci.* 39 (1): 1–15.

———. 1991. Sailing downwind: Aerodynamic performance of the *Velella* sail. *J. Exp. Biol.* 158: 117–132.

Francour, Patrice. 1997. Predation on holothurians: A literature review. *Invertebr. Biol.* 116 (1): 52–60.

Frank, Peter W. 1982. Effects of winter feeding on limpets by black oystercatchers, *Haematopus bachmani. Ecology* 63 (5): 1352–1362.

Gansel, John A. 1979. The symbiotic relationship between *Notoacmea paleacea* and *Phyllospadix torreyi:* Distribution by exposure and feeding behavior. M.S. thesis, Hopkins Marine Station, Stanford University, Pacific Grove, CA.

Garrison, David L., S. M. Conrad, P. P. Eilers, and E. M. Waldron. 1992. Confirmation of domoic acid production by *Pseudonitzschia australis* (Bacillariophyceae) cultures. *J. Phycol.* 28: 604–607.

Gerhart, Donald J., Daniel Rittschof, and Sara W. Mayo. 1988. Chemical ecology and the search for marine antifoulants: Studies of a predator-prey symbiosis. *J. Chem. Ecol.* 14: 1905–1917.

Gerhart, John, and Marc Kirschner. 1997. *Cells, embryos, and evolution: Toward a cellular and developmental understanding of phenotypic variation and evolutionary adaptability.* Cambridge, MA: Blackwell Sciences.

Hadfield, M. G., and C. N. Hopper. 1980. Ecological and evolutionary significance of pelagic spermatophores of vermetid gastropods. *Mar. Biol.* 57: 315–325.

Hadfield, Michael G., and Donna K. Iaea. 1989. Velum of encapsulated veligers of *Petaloconchus* (Gastropoda) and the problem of re-evolution of planktotrophic larvae. *Bull. Mar. Sci.* 45 (2): 377–386.

Hahn, Thomas, and Mark Denny. 1989. Tenacity-mediated selective predation by oystercatchers on intertidal limpets and its role in maintaining habitat partitioning by *"Collisella" scabra* and *Lottia digitalis. Mar. Ecol. Prog. Ser.* 53: 1–10.

Halanych, K. M., John D. Bacheller, Anna Marie A. Aguinaldo, Stephanie M. Liva, David M. Hillis, and James A. Lake. 1995. Evidence from 18S ribosomal DNA that the Lophophorates are protostome animals. *Science* 267: 1641–1642.

Halder, Georg, Patrick Callaerts, and Walter Gehring. 1995. Induction of ectopic eyes by targeted expression of the *eyeless* gene in *Drosophila. Science* 267: 1788–1792.

Hallacher, Leon E., and Dale A. Roberts. 1985. Differential utilization of space and food by the inshore rockfishes (Scorpaenidae: Sebastes) of Carmel Bay, California. *Environ. Biol. Fishes* 12 (2): 19–110.

Halliday, E. B. 1988. An unusual settlement of the barnacle *Megabalanus californicus* on turban snails *Tegula* spp. in a central California USA kelp forest. *Am. Zool.* 28 (4): 190A.

Hanlon, Roger T., and John B. Messenger. 1996. *Cephalopod behaviour*. Cambridge: Cambridge University Press.

Harlin, Marilyn M. 1973. "Obligate" algal epiphyte: *Smithora naiadum* grows on a synthetic substrate. *J. Phycol.* 9: 230–232.

Harris, Larry G. 1987. Aeolid nudibranchs as predators and prey. *Am. Malacol. Bull.* 5 (2): 287–292.

Hartney, K. B. 1996. Site fidelity and homing behaviour of some kelp-bed fishes. *J. Fish. Biol.* 49 (6): 1062–1069.

Hartwick, E. B. 1976. Foraging strategy of the black oyster catcher (*Haematopus bachmani Audubon*). *Canad. J. Zool.* 54: 142–155.

Harvell, C. Drew. 1984. Predator-induced defense in a marine bryozoan. *Science* 224: 1357–1359.

———. 1990. The ecology and evolution of inducible defenses. *Q. Rev. Biol.* 65 (3): 323–340.

———. 1998. Genetic variation and polymorphism in the inducible spines of a marine bryozoan. *Evolution* 52 (1): 80–86.

Haven, Stoner B. 1973. Competition for food between the intertidal gastropods *Acmaea scabra* and *Acmaea digitalis. Ecology* 54: 143–151.

Hawkes, Michael W. 1988. Evidence of sexual reproduction in *Smithora naiadum* (Erythropeltidales: Rhodophyta) and its evolutionary significance. *Br. Phycol. J.* 23: 327–336.

Hayward, P. J., and J. S. Ryland. 1985. *Cyclostome bryozoans.* London: Linnean Society; Estuarine and Brackish-Water Sciences Assoc.

Herrlinger, Timothy J., S. C. Schroeter, and J. Dixon. 1987. Giant kelp blades: An asteroid nursery ground? Western Society of Naturalists, annual meeting abstracts.

Himmelman, John H. 1975. Phytoplankton as a stimulus for spawning in three marine invertebrates. *J. Exp. Mar. Biol. Ecol.* 20: 199–214.

Hivnor, Maggie. 1986. Coaxing the fossil record back to life. *Univ. Chicago mag.* 78 (3): 2–9.

Hobson, E. S., and J. R. Chess. 1988. Trophic relations of the blue rockfish, *Sebastes mystinus*, in a coastal upwelling system off northern California (USA). *US Natl. Mar. Fish. Ser. Fish. Bull.* 86 (4): 715–743.

Hobson, Edmund S., James R. Chess, and Daniel F. Howard. 1996. Zooplankters consumed by blue rockfish during brief access to a current off California's Sonoma coast. *Calif. Dep. Fish Game Bull.* 82 (2): 87–92.

Holstein, T., and P. Tardent. 1984. An ultrahigh-speed analysis of exocytosis: Nematocyst discharge. *Science* 223: 830–832.

Holts, L. J., and K. A. Beauchamp. 1993. Sexual reproduction in the corallimorpharian sea anemone *Corynactis californica* in a central California kelp forest. *Mar. Biol.* 116: 129–136.
Howard, L. D. 1973. Muscular anatomy of the fore-limb of the sea otter (*Enhydra lutris*). *Proc. Calif. Acad. Sci.*, 4th ser., 39 (20): 411–500.
Howe, Nathan R. 1976. Proline inhibition of a sea anemone alarm pheromone response. *J. Exp. Biol.* 65: 147–156.
Howe, Nathan R., and Younus M. Sheikh. 1975. Anthopleurine: A sea anemone alarm pheromone. *Science* 189: 386–388.
Hyman, Libbie Henrietta. 1922. *Comparative vertebrate anatomy.* Chicago: University of Chicago Press.
———. 1940. *The invertebrates: Protozoa through Ctenophora.* New York: McGraw-Hill.
Jacobson, L. D., and R. D. Vetter. 1996. Bathymetric demography and niche separation of thornyhead rockfish: *Sebastolobus alascanus* and *Sebastolobus altivelis. Can. J. Fish Aquat. Sci.* 53 (3): 600–609.
Jannasch, Holger W., and Michael J. Mottl. 1985. Geomicrobiology of deep-sea hydrothermal vents. *Science* 229: 717–725.
Jensen, G. C. 1995. *Pacific coast crabs and shrimps.* Monterey, CA: Sea Challengers.
Johnson, W. S., Andreas Gigon, S. L. Gulmon, and H. A. Mooney. 1974. Comparative photosynthetic capacities of intertidal algae under exposed and submerged conditions. *Ecology* 55: 450–453.
Kain (Jones), Joanna M., and Trevor A. Norton. 1987. Growth of blades of *Nereocystis luetkeana* (Phaeophyta) in darkness. *J. Phycol.* 23: 464–469.
Kandel, Eric R. 1979. *Behavioral biology of Aplysia.* San Francisco: W. H. Freeman.
Kannan, Kurunthachalam, Keerthi S. Gurunge, Nancy J. Thomas, Shinsuke Tanabe, and John P. Giesy. 1998. Butyltin residues in southern sea otters (*Enhydra lutris nereis*) found dead along California coastal waters. *Environ. Sci. Technol.* 32: 1169–1175.
Karp, Richard D., and W. H. Hildemann. 1976. Specific allograft reactivity in the seastar *Dermasterias imbricata. Transplantation* (Baltimore) 22 (5): 434–439.
Kartner, Norbert, John R. Riordan, and Victor Ling. 1983. Cell surface P-glycoprotein associated with multidrug resistance in mammalian cell lines. *Science* 221: 1285–1288.
Kaufman, Melissa R., Yuzuru Ikeda, Chris Patton, Gilbert Van Dykhuizen, and David Epel. 1998. Bacterial symbionts colonize the accessory nidamental gland of the squid *Loligo opalescens* via horizontal transmission. *Biol. Bull.* 194: 36–43.
Kenyon, Karl W. 1975. *The sea otter in the eastern Pacific ocean.* New York: Dover.
Keough, Michael J. 1984. Kin-recognition and the spatial distribution of larvae of the bryozoan *Bugula neritina* (L.). *Evolution* 38 (1): 142–147.
Kohn, Alan J. 1966. Food specialization in *Conus* in Hawaii and California. *Ecology* 47 (6): 1041–1043.
Kornmann, Peter. 1938. Zur Entwicklungsgeschichte von *Derbesia* und *Halicystis. Planta* 28: 464–470.
Kozloff, Eugene N. 1983. *Seashore life of the northern Pacific coast.* Seattle: University of Washington Press.
Kurelec, Branko, and Branka Pivčević. 1992. The multidrug resistance-like mechanism in the marine sponge *Tethya aurantium. Mar. Environ. Res.* 34: 249–253.
Kuwazuru, Y., A. Yoshimura, S. Hanada, A. Utsunomiya, T. Makino, K. Ishibashi, M. Kodama, M. Iwahashi, T. Arima, and S.-I. Akiyama. 1990. Expression of the multidrug transporter, P-glycoprotein, in acute leukemia cells and correlation to clinical drug resistance. *Cancer* 66 (5): 868–873.
Kvitek, Rikk Glenn, C. Bretz, and K. Thomas. 1999. Influence of harmful algal bloom toxins on the distribution and foraging behavior of sea otters and shorebirds. Western Society of Naturalists, annual meeting abstracts.

Kvitek, Rikk Glenn, A. R. DeGange, and M. K. Beitler. 1991. Paralytic shellfish poisoning toxins mediate feeding behavior of sea otters. *Limnol. Oceanogr.* 36 (2): 393–404.

LaBarbera, M. 1981. Water flow patterns in and around three species of articulate brachiopods. *J. Exp. Mar. Biol. Ecol.* 55: 185–206.

Lambert, C. C., and G. Lambert. 1998. Non-indigenous ascidians in southern California harbors and marinas. *Mar. Biol.* 130: 675–688.

Land, M. F. 1980. Compound eyes: Old and new optical mechanisms. *Nature* 287: 681–686.

Langdon, S. C. 1917. Carbon monoxide, occurrence free in kelp (*Nereocystis luetkeana*). *J. Am. Chem. Soc.* 39: 149–156.

Larson, Ronald J. 1980. The medusa of *Velella velella* (Linnaeus, 1758) (Hydrozoa: Chondrophorae). *J. Plankton Res.* 2 (3): 183–186.

Lawrence, J. M. 1990. A chemical alarm response in *Pycnopodia helianthoides* (Echinodermata: Asteroidea). *Am. Zool.* 30 (4): 117A.

Lee, Robert Edward. 1980. *Phycology.* Cambridge: Cambridge University Press.

Lee, Welton L. 1966. Pigmentation of the marine isopod *Idothea montereyensis. Comp. Biochem. Physiol.* 18: 17–36.

———. 1972. Chromatophores and their rôle in color change in the marine isopod *Idotea montereyensis* (Maloney). *J. Exp. Mar. Bio. Ecol.* 8: 201–215.

Lee, W. L., and B. M. Gilchrist. 1972. Pigmentation, color change, and the ecology of the marine isopod *Idotea resecata* (Stimpson, 1857). *J. Exp. Mar. Bio. Ecol.* 10: 1–27.

Lewis, Cindy A. 1978. A review of substratum selection in free-living and symbiotic cirripeds. In *Settlement and metamorphosis of marine invertebrate larvae*, ed. F. S. Chia and M. E. Rice, 207–218. New York: Elsevier.

Li, Chia-Wei, Jun-Yuan Chen, and Tzu-En Hua. 1998. Precambrian sponges with cellular structures. *Science* 279: 879–882.

Lindquist, Niels, and Mark E. Hay. 1996. Palatability and chemical defense of marine invertebrate larvae. *Ecol. Monogr.* 66 (4): 431–450.

Lindsay, S. M., T. M. Frank, J. Kent, J. C. Partridge, and M. I. Latz. 1999. Spectral sensitivity of vision and bioluminescence in the midwater shrimp *Sergestes similis. Biol. Bull.* 197: 348–360.

Lindstedt, K. June. 1971. Biphasic feeding response in a sea anemone: Control by asparagine and glutathione. *Science* 173: 333–334.

Lisin, S., J. P. Barry, and C. Harrold. 1993. Reproductive biology of vesicomyid clams from cold seeps in Monterey Bay, California. *Am. Zool.* 33 (5): 50A.

Lobban, Christopher S., and Paul J. Harrison. 1994. *Seaweed ecology and physiology*. Cambridge: Cambridge University Press.

Love, Robin Milton. 1991. *Probably more than you want to know about the fishes of the Pacific Coast.* Santa Barbara, CA: Really Big Press.

Lowe, C. G., R. N. Bray, and D. R. Nelson. 1994. Feeding and associated electrical behavior of the Pacific electric ray *Torpedo californica* in the field. *Mar. Biol.* 120 (1): 161–169.

Lubbock, Roger. 1980. Clone-specific cellular recognition in a sea anemone. *Proc. Natl. Acad. Sci. USA* 77: 6667–6669.

Lubbock, Roger, and G. A. B. Shelton. 1981. Electrical activity following cellular recognition of self and non-self in a sea anemone. *Nature* 289: 59–60.

Lucero, Mary T., Heraldo Farrington, and William F. Gilly. 1994. Quantification of L-dopa and dopamine in squid ink: Implications for chemoreception. *Biol. Bull.* 187: 55–63.

Lyons, Kathy J. 1986. New studies reveal sea otters are highly specialized foragers. *Otter Raft* (newsletter of Friends of the Sea Otter, Carmel, Calif.), summer: 11.

MacGinitie, G. E. 1934. The egg-laying activities of the sea hare, *Tethys californicus* (Cooper). *Biol. Bull.* (Woods Hole), 67: 300–303.

MacGinitie, G. E., and Nettie MacGinitie. 1968. *Natural history of marine animals.* 2d ed., rev. New York: McGraw-Hill.
Madin, L. P. 1974. Field observations of the feeding behavior of salps (Tunicata: Thaliacea). *Mar. Biol.* 25: 143–147.
———. 1990. Aspects of jet propulsion in salps. *Canad. J. Zool.* 68: 765–777.
Maier, I., and D. G. Müller. 1986. Sexual pheromones in algae. *Biol. Bull.* 170: 145–175.
Main, Kevan L. 1982. The early development of two ovulid snails, *Simnia aequalis* and *barbarensis. Veliger* 24 (3): 252–258.
Mann, Charles C. 1999. Genetic engineers aim to soup up crop photosynthesis. *Science* 283: 314–316.
Margolin, Abe S. 1964. The mantle response of *Diodora aspera. Anim. Behav.* 12 (1): 187–194.
Mariscal, Richard N., Edwin J. Conklin, and Charles H. Bigger. 1977. The ptychocyst, a major new category of cnida used in tube construction by a cerianthid anemone. *Biol. Bull.* 152: 392–405.
Marsden, Joan Rattenbury. 1957. Regeneration in *Phoronis vancouverensis. J. Morphol.* 101: 307–323.
Matzel, L. D. 1990. Pavlovian conditioning of distinct components of *Hermissenda*'s responses to rotation. *Behav. Neural. Biol.* 54 (2): 131–145.
Mauch, Shona, and Joel Elliott. 1997. Protection of the nudibranch *Aeolidia papillosa* from nematocyst discharge of the sea anemone *Anthopleura elegantissima. Veliger* 40 (2): 148–151.
McFadden, C. S., R. K. Grosberg, B. B. Cameron, D. P. Karlton, and D. Secord. 1997. Genetic relationships within and between clonal and solitary forms of the sea anemone *Anthopleura elegantissima* revisited: Evidence for the existence of two species. *Mar. Biol.* 128: 127–139.
McGinnis, William, and Michael Kuziora. 1994. The molecular architects of body design. *Sci. Am.*, Feb. 58–66.
McLaughlin, P. A., and G. C. Jensen. 1996. A new species of hermit crab of the genus *Parapagurodes* (Decapoda: Anomura; Paguridae) from the eastern Pacific, with a description of its first zoeal stage. *J. Nat. Hist.* 30: 841–854.
Meese, R. J. 1993. Effects of predation by birds on gooseneck barnacle *Pollicipes polymerus* Sowerby distribution and abundance. *J. Exp. Mar. Biol. Ecol.* 166 (1): 47–64.
Miller, Linda J. 1998. Apoptosis. *Science* 281: 1301.
Mills, C. E. 1981. Diversity of swimming behaviors in hydromedusae as related to feeding and utilization of space. *Mar. Biol.* 64: 185–189.
———. 1994. Seasonal swimming of sexually mature benthic opisthobranch molluscs (*Melibe leonina* and *Gastropteron pacificum*) may augment population dispersal. In *Reproduction and development of marine invertebrates: Symposium, Friday Harbor, WA, June 9–11, 1992,* ed. W. H. Wilson, S. A. Stricker, and G. L. Shinn, 313–319. Baltimore: Johns Hopkins University Press.
———. 1997. Polyorchid hydromedusae may be disappearing from bays and inlets on both east and west coasts of the north Pacific ocean. Western Society of Naturalists, annual meeting abstracts.
Milne, Lorus J., and Margery Milne. 1967. *Patterns of survival.* Englewood Cliffs, NJ: Prentice-Hall.
Mladenov, P. V., and F. S. Chia. 1983. Development, settling behaviour, metamorphosis, and pentacrinoid feeding and growth of the feather star *Florometra serratissima. Mar. Biol.* 73: 309–323.
Moitoza, D. J., and D. W. Phillips. 1979. Prey defense, predator preference, and nonrandom diet: The interactions between *Pycnopodia helianthoides* and two species of sea urchins. *Mar. Biol.* 53: 299–304.
Molenock, Joane, and Edgardo D. Gomez. 1972. Larval stages and settlements of the barnacle *Balanus (Conopea) Galeatus* (L.) (*Cirripedia Thoracica*). *Crustaceana* 23 (1): 100–108.
Monniot, Claude, Françoise Monniot, and Pierre Laboute. 1991. *Coral reef ascidians of New Caledonia.* Editions de l'ORSTOM, Institut français de recherche scientifique pour le développement en coopération, Collection Faune tropicale no. 30. Paris.
Morris, Robert H., Donald P. Abbott, and Eugene C. Haderlie, eds. 1980. *Intertidal invertebrates of California.* Stanford: Stanford University Press.

Morse, Daniel E. 1990. Recent progress in larval settlement and metamorphosis: Closing the gaps between molecular biology and ecology. *Bull. Mar. Sci.* 46 (2): 465–483.

Mukai, Hideo, and Hiroshi Watanabe. 1975. Fusibility of colonies in natural populations of the compound ascidian *Botrylloides violaceus. Proc. Jpn. Acad.* 51: 48–50.

Müller, W. E. G., R. Steffen, B. Rinkevich, V. Matranga, and B. Kurelec. 1996. The multixenobiotic resistance mechanism in the marine sponge *Suberites domuncula:* Its potential applicability for the evaluation of environmental pollution by toxic compounds. *Mar. Biol.* 125: 165–170.

Murphy, Philip G. 1978. *Collisella austrodigitalis* sp. nov.: A sibling species of limpet (Acmaeidae) discovered by electrophoresis. *Biol. Bull.* 155: 293–206.

Muscatine, Leonard, and Cadet Hand. 1958. Direct evidence for the transfer of materials from symbiotic algae to the tissues of a coelenterate. *Proc. Natl. Acad. Sci. USA* 44: 1259–1263.

Nelson, J. Lee. 1998. Microchimerism and autoimmune disease. *N. Eng. J. Med.* 338 (17): 1224–1225.

Nelson, Thomas J., Carlos Collin, and Daniel L. Alkon. 1990. Isolation of a G protein that is modified by learning and reduces potassium currents in *Hermissenda. Science* 247: 1479–1483.

Newman, William A., and Donald P. Abbott. 1980. Cirripedia: The barnacles. In *Intertidal invertebrates of California,* ed. Robert H. Morris, Donald P. Abbott, and Eugene C. Haderlie, 504–535. Stanford: Stanford University Press.

Newman, William A., and Ronald R. McConnaughey. 1987. A tropical eastern Pacific barnacle, *Megabalanus coccopoma* (Darwin), in southern California, following El Niño 1982–83. *Pac. Sci.* 41 (1–4): 31–36.

Nicotri, M. E. 1977. Grazing effects of four marine intertidal herbivores on the microflora. *Ecology* 58: 1020–1032.

Nielsen, Claus. 1998. Origin and evolution of animal life cycles. *Biol. Rev.* 73: 125–155.

Nilsson, Dan E. 1994. Eyes as optical alarm systems in fan worms and ark clams. *Philos. Trans. R. Soc. Lond. B Biol. Sci.* 346: 195–212.

Nolen, T. G., P. M. Johnson, C. E. Kicklighter, and T. Capo. 1995. Ink secretion by the marine snail *Aplysia californica* enhances its ability to escape from a natural predator. *J. Comp. Physiol. A* 176: 239–254.

Norman, J. R. 1975. *A history of fishes,* 3d ed., by P. H. Greenwood. London: Ernest Benn.

Norton, Stephen F. 1988. Role of the gastropod shell and operculum in inhibiting predation by fishes. *Science* 241: 92–94.

Nutting, Charles C. 1909. Alcyonaria of the California coast. *Proc. US Natl. Mus.* 35 (1658): 681–688.

Nybakken, James W. 1997. *Marine biology: An ecological approach.* 4th ed. Reading, MA: Addison Wesley Longman.

O'dor, R. K. 1988. The forces acting on swimming squid. *J. Exp. Biol.* 137: 421–442.

Oehlmann, J., E. Stroben, and P. Fioroni. 1993. Frequency and degree of expression of pseudohermaphroditism in some stenoglossan prosobranchs from the French coasts (especially the bay of Morlaix and the English Channel). *Cah. Biol. Mar.* 34 (3): 343–362.

Ostarello, Georgiandra Little. 1973. Natural history of the hydrocoral *Allopora californica,* Verrill (1866). *Biol. Bull.* 145: 548–564.

———. 1976. Larval dispersal in the subtidal hydrocoral *Allopora californica,* Verrill (1866). In *Coelenterate ecology and behavior,* ed. G. O. Mackie, 331–337. New York: Plenum.

O'Sullivan, J. B., R. R. McConnaughey, and M. E. Huber. 1987. A blood-sucking snail: The Cooper's nutmeg, *Cancellaria cooperi,* Gabb, parasitizes the California electric ray, *Torpedo californica,* Ayers. *Biol. Bull.* 172 (3): 362–366.

Painter, Sherry D., Bret Clough, Rebecca W. Garden, Jonathan V. Sweedler, and Gregg T. Nagle. 1998. Characterization of *Aplysia* attractin, the first water-borne peptide pheromone in invertebrates. *Biol. Bull.* 194: 120–131.

Panganiban, Grace, Steven M. Irvin, Chris Lowe, Henry Roehl, Laura S. Corley, Beverley Sherbon, Jennifer K. Grenier, John F. Fallon, Judith P. Kimble, Muriel Walker, Gregory A. Wray, Billie J. Swalla, Mark Q. Martindale, and Sean B. Carroll. 1997. The origin and evolution of animal appendages. *Proc. Natl. Acad. Sci.* 94: 5162–5166.

Parker, B. C. 1971. Studies of translocation in *Macrocystis.* In *The biology of giant kelp beds (Macrocystis) in California,* ed. Wheeler J. North, 191–195. Beihefte zur Nova Hedwigia, no. 32. Lehre: J. Cramer.

Patella, V., V. Casolaro, A. Ciccarelli, G. R. Pettit, M. Columbo, and G. Marone. 1995. The antineoplastic bryostatins affect human basophils and mast cells differently. *Blood* 85 (5): 1272–1281.

Patton, Wendell K. 1972. Studies on the animal symbionts of the gorgonian coral, *Leptogorgia virgulata* (Lamarck). *Bull. Mar. Sci.* 22 (2): 419–431.

Pawlik, Joseph R., John B. O'Sullivan, and M. G. Harasewych. 1988. The egg capsules, embryos, and larvae of *Cancellaria cooperi* (Gastropoda: Cancellariidae). *Nautilus* 102 (2): 47–53.

Pearse, V., J. Pearse, M. Buchsbaum, and R. Buchsbaum. 1987. *Living invertebrates.* Palo Alto, CA: Blackwell Scientific.

Peck, L. S. 1993. The tissues of articulate brachiopods and their value to predators. *Philos. Trans. R. Soc. Lond. B Biol. Sci.* 339 (1287): 17–32.

Pennings, Steven C. 1990a. Multiple factors promoting narrow host range in the sea hare, *Aplysia californica. Oecologia* 82: 192–200.

———. 1990b. Size-related shifts in herbivory: Specialization in the sea hare *Aplysia californica* Cooper. *J. Exp. Mar. Biol. Ecol.* 142: 43–61.

———. 1991. Reproductive behavior of *Aplysia californica* Cooper: Diel patterns, sexual roles, and mating aggregations. *J. Exp. Mar. Biol. Ecol.* 149: 249–266.

Pennington, J. T., M. N. Tamburri, and J. P. Barry. 1997. *Laqueus californianus* (Brachiopoda): Larval temperature tolerance and settlement preference. Western Society of Naturalists, annual meeting abstracts.

Pettitt, John, Sophie Ducker, and Bruce Knox. 1981. Submarine pollination. *Sci. Am.,* March: 135–143.

Phillips, D. W. 1978. Chemical mediation of invertebrate defensive behaviors and the ability to distinguish between foraging and inactive predators. *Mar. Biol.* 49: 237–243.

Phillips, Julius B. 1964. Life history studies on ten species of rock fish (genus *Sebastodes*). *Calif. Dep. Fish Game Fish. Bull.* 126: 5–39.

Polanshek, Alan R., and John A. West. 1977. Culture and hybridization studies on *Gigartina papillata* (Rhodophyta). *J. Phycology* 13: 141–149.

Pomeroy, Lawrence R. 1974. The ocean's food web, a changing paradigm. *Bioscience* 24 (9): 499–504.

Postels, Alexandre. 1836. *Voyage autour du monde, exécuté par ordre de sa Majesté l'Empereur Nicolas Ier par F. Lutké. Tome troisième contenant les travaux de MM. les naturalistes.* Paris: Firmin Didot Frères.

Punnett, Thomas, Richard L. Miller, and Bong-Hee Yoo. 1992. Partial purification and some chemical properties of the sperm chemoattractant from the forcipulate starfish *Pycnopodia helianthoides* (Brandt 1835). *J. Exp. Zool.* 262: 87–96.

Putnam, Deborah A. 1964. The dispersal of young of the commensal gastropod *Crepidula adunca* from its host, *Tegula funebralis. Veliger* 6 suppl.: 63–66.

Reed, D. C. 1990. The effects of variable settlement and early competition on patterns of kelp recruitment. *Ecology* 71 (2): 776–787.

Reed, Daniel C., Charles D. Amsler, and Alfred W. Ebeling. 1992. Dispersal in kelps: Factors affecting spore swimming and competency. *Ecology* 73 (5): 1577–1585.

Reed, Daniel C., Todd W. Anderson, Alfred W. Ebeling, and Michele Anghera. 1997. The role of reproductive synchrony in the colonization potential of kelp. *Ecology* 78 (8): 2443–2457.

Reid, R. G. B., and B. D. Gustafson. 1989. Update on feeding and digestion of the moon snail *Polinices lewisii* (Gould 1847). *Veliger* 32 (3): 327.

Ricketts, Edward F., Jack Calvin, and Joel W. Hedgpeth. Revised by David W. Phillips. 1985. *Between Pacific tides*. 5th ed. Stanford: Stanford University Press.

Riise, Scott A. 1990. The distribution, feeding behavior, and anthocodia retraction of *Anthomastus ritteri* Nutting, 1909, in Monterey Bay. MS., Hopkins Marine Station, Stanford University, Pacific Grove, CA.

Rinkevich, Baruch, Robert J. Lauzon, Byron W. M. Brown, and Irving L. Weissman. 1992. Evidence for a programmed life span in a colonial protochordate. *Proc. Natl. Acad. Sci. USA* 89: 3546–3550.

Rinkevich, Baruch, Tami Lilker Levav, and Menachem Goren. 1994. Allorecognition/xenorecognition responses in *Botrylloides* (Ascidiacea) subpopulations from the Mediterranean coast of Israel. *J. Exp. Zool.* 270: 302–312.

Rittschof, Dan. 1980. Chemical attraction of hermit crabs and other attendants to simulated gastropod predation sites. *J. Chem. Ecol.* 6 (1): 103–118.

Robison, Bruce H. 1995. Light in the ocean's midwaters. *Sci. Am.*, July: 60–64.

Rosenthal, Lara. 1989. Insulation strategies in the Pacific electric ray *Torpedo californica*. MS., Hopkins Marine Station, Stanford University, Pacific Grove, CA.

Russel, D. J., and J. W. Hedgpeth. 1990. Host utilization during ontogeny by two pycnogonid species (*Tanystylum duospinum* and *Ammothea hilgendorfi*) parasitic on the hydroid *Eucopella everta* (Coelenterata: Campanulariidae). *Bijdr. Dierkd.* 60 (3–4): 215–224.

Rutowski, Ronald L. 1983. Mating and egg mass production in the aeolid nudibranch *Hermissenda crassicornis* (Gastropoda: Opisthobranchia). *Biol. Bull.* 165: 276–285.

Saffo, Mary Beth. 1987. New light on seaweeds. *BioScience* 37 (9): 654–664.

Saito, Yasunori, Euichi Hirose, and Hiroshi Watanabe. 1994. Allorecognition in compound ascidians. *Int. J. Dev. Biol.* 38: 237–247.

Salleo, Alberto, Giovanni Musci, Paolo F. A. Barra, and Lilia Calabrese. 1996. The discharge mechanism of acontial nematocytes involves the release of nitric oxide. *J. Exp. Biol.* 199: 1261–1267.

Sanders, B. M., C. Hope, V. M. Pascoe, and L. S. Martin. 1991. Characterization of the stress protein response in two species of *Collisella* limpets with different temperature tolerances. *Physiol. Zool.* 64 (6): 1471–1489.

Scheller, Richard H., Rashad-Rudolf Kaldany, Thane Kreiner, Anne C. Mahon, John R. Nambu, Mark Schaefer, and Ronald Taussig. 1984. Neuropeptides: Mediators of behavior in *Aplysia*. *Science* 225: 1300–1308.

Scholin, Christopher A., Frances Gulland, Gregory J. Doucette, and 23 others. 2000. Mortality of sea lions along the Central California coast linked to a toxic diatom bloom. *Nature* 403: 80–84.

Scofield, Virginia L., Jay M. Schlumpberger, and Irving Weissman. 1982. Colony specificity in the colonial tunicate *Botryllus* and the origins of vertebrate immunity. *Amer. Zool.* 22: 783–794.

Seed, R. 1976. Observations on the ecology of *Membranipora* (Bryozoa) and a major predator *Doridella steinbergae* (Nudibranchiata) along the fronds of *Laminaria saccharina* at Friday Harbor, Washington. *J. Exp. Mar. Biol. Ecol.* 24: 1–17.

Seiff, Stuart R. 1975. Predation upon subtidal *Tonicella lineata* of Mussel Point, California. In *The biology of Chitons: Veliger* 18 suppl., July 15.

Serrão, Ester A., Gareth Pearson, Lena Kautsky, and Susan H. Brawley. 1996. Successful external fertilization in turbulent environments. *Proc. Natl. Acad. Sci. USA* 93: 5286–5290.

Sewell, Mary A., and Don R. Levitan. 1992. Fertilization success during a natural spawning of the dendrochirote sea cucumber *Cucumaria miniata*. *Bull. Mar. Sci.* 51 (2): 161–166.

Shaw, G. D., and A. R. Fontaine. 1990. The locomotion of the comatulid *Florometra serratissima* (Echinodermata: Crinoidea) and its adaptive significance. *Can. J. Zool.* 68: 942–950.

Shears, Margaret A., Ming H. Kao, Gary K. Scott, Peter L. Davies, and Garth L. Fletcher. 1993. Distribution of type III antifreeze proteins in the Zoarcoidei. *Mol. Mar. Biol. Biotechnol.* 2 (2): 104–111.

Shenker, J. M. 1988. Oceanographic associations of neustonic larval and juvenile fishes and dungeness crab megalopae off Oregon (USA). *US Natl. Mar. Fish. Serv. Fish. Bull.* 86 (2): 299–318.

Siebenaller, Joseph H. 1984. Pressure-adaptive differences in NAD-dependent dehydrogenases of congeneric marine fishes living at different depths. *J. Comp Physiol. B* 154: 443–448.

Simpson, Tracy L. 1984. *The cell biology of sponges.* New York: Springer.

Singer, Susan R. 1997. Plant life cycles and angiosperm development. In *Embryology: Constructing the organism,* ed. Scott F. Gilbert and Anne M. Raunio, 493–513. Sunderland, MA: Sinauer Assoc.

Slocum, Carol J. 1980. Differential susceptibility to grazers in two phases of an intertidal alga: Advantages of heteromorphic generations. *J. Exp. Mar. Biol. Ecol.* 46: 99–110.

Smith, A. M. 1991. The role of suction in the adhesion of limpets. *J. Exp. Biol.* 161: 151–169.

Smith, John R., and David R. Strehlow. 1983. Algal-induced spawning in the marine mussel *Mytilus californianus. Int. J. Invertebr. Reprod.* 6: 129–133.

Smith, K. L., Jr., and N. O. Brown. 1983. Oxygen consumption of pelagic juveniles and demersal adults of the deep-sea fish *Sebastolobus altivelis,* measured at depth. *Mar. Biol.* 76: 325–332.

Smith, Robert H. 1971. Reproductive biology of a brooding sea-star *Leptasterias pusilla* (Fisher), in the Monterey bay region. Ph.D. diss., Stanford University.

Snyder, N., and H. Snyder. 1970. Alarm response of *Diadema antillarum. Science* 168: 276–278.

Sommer, F. A. 1988. Developmental cycle of *Pelagia colorata* (Scyphozoa: Pelagiidae). *Am. Zool.* 28: 170A, no. 897.

Standing, Jon D., I. R. Hooper, and J. D. Costlow. 1984. Inhibition and induction of barnacle settlement by natural products present in octocorals. *J. Chem. Ecol.* 10 (6): 823–834.

Starr, Michel, John H. Himmelman, and Jean-Claude Therriault. 1990. Direct coupling of marine invertebrate spawning with phytoplankton blooms. *Science* 247: 1071–1074.

Stephenson, T. A., and A. Stephenson. 1949. The universal features of zonation between tide-marks on rocky coasts. *J. Ecol.* 37: 289–305.

Stepien, Carol A., Marlen Glattke, and Keith M. Fink. 1988. Regulation and significance of color patterns of the spotted kelpfish, *Gibbonsia elegans,* Cooper, 1864 (Blennioidei: Clinidae). *Copeia* 1988 (1): 7–15.

Stoner, Douglas S., and Irving Weissman. 1996. Somatic and germ cell parasitism in a colonial ascidian: Possible role for a highly polymorphic allorecognition system. *Proc. Natl. Acad. Sci. USA* 93: 15254–15259.

Strand, S. W., and W. M. Hammer. 1988. Predatory behavior of *Phacellophora camtschatica* and size-selective predation upon *Aurelia aurita* (Scyphozoa: Cnidaria) at Saanich Inlet, British Columbia. *Mar. Biol.* 99: 409–414.

Sullivan, K. M., and G. N. Somero. 1980. Enzyme activities of fish skeletal muscle and brain as influenced by depth of occurrence and habits of feeding and locomotion. *Mar. Biol.* 60: 91–99.

Suttle, Curtis A. 1999. Do viruses control the oceans? *Nat. Hist.* 108 (1): 48–51.

Tam, Y. K., I. Kornfield, and F. P. Ojeda. 1996. Divergence and zoogeography of mole crabs, *Emerita* spp. (Decapoda: Hippidae), in the Americas. *Mar. Biol.* (Berlin) 125 (3): 489–497.

Tegner, Mia J., and Paul K. Dayton. 1977. Sea urchin recruitment patterns and implications of commercial fishing. *Science* 196: 324–326.

Teitelbaum, Jeanne S., R. J. Zatorre, S. Carpenter, D. Gendron, A. C. Evans, A. Gjedde, and N. R. Cashman. 1990. Neurologic sequelae of domoic acid intoxication due to the ingestion of contaminated mussels. *New Engl. J. Med.* 322 (25): 1781–1787.

Temkin, M. H. 1994. Gamete spawning and fertilization in the Gymnolaemate bryozoan *Membranipora membranacea. Biol. Bull.* 187: 143–155.

Terrados, J., and S. L. Williams. 1997. Leaf versus root nitrogen uptake by the surfgrass *Phyllospadix torreyi. Mar. Ecol. Prog. Ser.* 149 (1–3): 267–277.

Terres, John K. 1980. *The Audubon Society encyclopedia of North American birds.* New York: Knopf.

Thayer, Charles W. 1985. Brachiopods versus mussels: Competition, predation, and palatability. *Science* 228: 1527–1528.

Theodor, Jacques. 1967. Contribution à l'étude des gorgones. VI. La dénudation des branches de gorgones par des mollusques prédateurs. *Vie et milieu* 18: 73–78.

———. 1970. Distinction between "self" and "not-self" in lower invertebrates. *Nature* 227: 690–692.

Thompson, D'Arcy Wentworth, trans. 1910. *The works of Aristotle.* Vol. 4: *Historia animalium.* Ed. J. A. Smith and W. D. Ross. Oxford: Clarendon Press.

Tomarev, Stanislav I., Patrick Callaerts, Lidia Kos, Rina Zinovieva, and George Halder. 1997. Squid Pax-6 and eye development. *Proc. Natl. Acad. Sci. USA* 94: 2421–2426.

Tomaschko, K.-H . 1995. Autoradiographic and morphological investigation of the defensive ecdysteroid glands in adult *Pycnogonum litorale* (Arthropoda: Pantopoda). *Eur. J. Entomol.* 92 (1): 105–112.

Topp, Werner, and R. A. Ring. 1988. Adaptations of coleoptera to the marine environment. I. Observations on rove beetles (Staphylinidae) from sandy beaches. *Can. J. Zool.* 66: 2464–2468.

Turner, Teresa. 1983. Facilitation as a successional mechanism in a rocky intertidal community. *Am. Nat.* 121 (5): 729–738.

Vasquez, J. A. 1993. Effects on the animal community of dislodgement of holdfasts of *Macrocystis pyrifera. Pac. Sci.* 47 (2): 180–184.

Verrill, A. E. 1922. *Canadian Arctic Expedition 1913–18,* vol. 8, pt. G. Ottawa: F. A. Acland.

Voltzow, Janice, and Rachel Collin. 1995. Flow through mantle cavities revisited: Was sanitation the key to fissurellid evolution? *Invert. Biol.* 114 (2): 145–150.

Vreeland, Valerie, John A. West, and Lynn Epstein. 1995. *Fucus* zygote attachment by formation of a vanadate bromoperoxidase-activated polyphenol glue. *J. Phycol.* 31 (3 suppl.): 4.

Wakefield, W. W., and K. L. Smith Jr. 1990. Ontogenetic vertical migration in *Sebastolobus altivelis* as a mechanism for transport of particulate organic matter at continental slope depths. *Limnol. Oceanogr.* 35 (6): 1314–1328.

Walker, Graham. 1995. Larval settlement: Historical and future perspectives. In *New frontiers in barnacle evolution,* ed. F. R. Schram and J. T. Høeg, 69–85. Crustacean Issues 10. Rotterdam and Brookfield, Vt.: A. A. Balkema.

Warner, G F., and J. D. Woodley 1975. Suspension-feeding in the brittle-star *Ophiothrix fragilis. J. Mar. Biol. Ass. UK* 55: 199–210.

Watanabe, James M., and Christopher Harrold. 1991. Destructive grazing by sea urchins, *Strongylocentrotus* spp., in a central California kelp forest: Potential roles of recruitment, depth, and predation. *Mar. Ecol. Prog. Ser.* 71: 125–141.

Watson, Glen M., and David A. Hessinger. 1989. Cnidocyte mechanoreceptors are tuned to the movements of swimming prey by chemoreceptors. *Science* 243: 1589–1591.

Watson, W. H., III, and C. M. Chester. 1993. The influence of olfactory and tactile stimuli on the feeding behavior of *Melibe leonina* (Gould, 1852) (Opisthobranchia: Dendronotacea). *Veliger* 36 (4): 311–316.

Wendell, F. 1994. Relationship between sea otter range expansion and red abalone abundance and size distribution in central California. *Calif. Dep. Fish Game Fish. Bull.* 80 (2): 45–56.

West, Hillary H. 1979. Pigmentation in the sea anemone *Corynactis californica. Comp. Biochem. Physiol.* 64B: 195–200.

Whittaker, J. R. 1997. Chordate evolution and autonomous specification of cell fate: The ascidian embryo model. *Amer. Zool.* 32: 237–249.

Wickelgren, Ingrid. 1998. Drug may suppress the craving for nicotine. *Science* 282: 1797–1799.
———. 1999. Filling in the blanks of the $GABA_B$ receptor. *Science* 283: 14–15.
Wicksten, Mary K. 1988. Antennary cast-net feeding in California hermit crabs (Decapoda: Paguridea). *Crustaceana* 54 (3): 321–322.
———. 1989. Why are there bright colors in sessile marine invertebrates? *Bull. Mar. Sci.* 45 (2): 519–530.
———. 1993. A review and a model of decorating behavior in spider crabs (Decapoda, Brachyura, Majidae). *Crustaceana* 64 (3): 314–325.
Widder, Edith A., Michael I. Latz, and James F. Case. 1983. Marine bioluminescence spectra measured with an optical multichannel detection system. *Biol. Bull.* 165: 791–810.
Willard, James M. 1981. The role of adaptive immunity in the predatory behavior of *Dermasterias imbricata*. M.S. thesis, California State University, Hayward.
Willcocks, Patricia Anne. 1980. Colonization and distribution of the epiphytes *Melobesia mediocris* and *Smithora naiadum* on the seagrass *Phyllospadix torreyi*. M.S. thesis, Hopkins Marine Station, Stanford University, Pacific Grove, CA.
Williams, Gary C. 1997. A new genus and species of nephtheid soft coral (Octocorallia: Alcyonacea) from the western Pacific Ocean, and a discussion of convergence with several deep-sea benthic organisms. *Proc. Calif. Acad. Sci.* 49 (12): 423–437.
Williams, S. L. 1995. Surfgrass (*Phyllospadix torreyi*) reproduction: Reproductive phenology, resource allocation, and male rarity. *Ecology* 76 (6): 1953–1970.
Wilson, E. O. 1992. *The diversity of life.* Cambridge, Mass.: Belknap Press of Harvard University Press.
Wilson, H. V. 1891. Notes on the development of some sponges. *J. Morphol.* 5: 511–519.
Winkler, Lindsay R., and Bernard E. Tilton. 1962. Predation on the California sea hare, *Aplysia californica* Cooper, by the solitary great green sea anemone, *Anthopleura xanthogrammica* (Brandt), and the effect of sea hare toxin and acetylcholine on anemone muscle. *Pac. Sci.* 16: 286–290.
Winston, Judith E. 1978. Polypide morphology and feeding behavior in marine ectoprocts. *Bull. Mar. Sci.* 28 (1): 1–31.
Wobber, Don R. 1970. A report on the feeding of *Dendronotis iris* on the anthozoan *Cerianthus* sp. from Monterey Bay, California. *Veliger* 12 (4): 383–387.
———. 1975. Agonism in asteroids. *Biol. Bull.* 148: 483–496.
Woll, C. 1997. Motility in juvenile gooseneck barnacles *Pollicipes polymerus*. Western Society of Naturalists, annual meeting abstracts.
Woltereck, R. 1904. Über die Entwicklung der Velella aus einer in der Tiefe vorkommenden Larve. In *Festschrift zum 70. Geburtstage des Herrn Geh. Raths Prof. Dr. A. Weisman, Zoologische Jahrbücher,* suppl. 7: 347–372. Jena: Gustav Fischer.
Wootton, J. Timothy. 1992. Indirect effects, prey susceptibility, and habitat selection: Impacts of birds on limpets and algae. *Ecology* 73 (3): 981–991.
———. 1994. Predicting direct and indirect effects: An integrated approach using experiments and path analysis. *Ecology* (Tempe) 75 (1): 151–165.
———. 1995. Effects of birds on sea urchins and algae: A lower-intertidal trophic cascade. *Ecoscience* 2 (4): 321–328.
Xiao, Shuhai, Yun Zhang, and Andrew H. Knoll. 1998. Three-dimensional preservation of algae and animal embryos in a Neoproterozoic phosphorite. *Nature* 391: 553–558.
Yoshiyama, R. M., W. D. Wallace, J. L. Burns, A. L. Knowlton, and J. R. Welter. 1996. Laboratory food choice by the mosshead sculpin, *Clinocottus globiceps* (Girard) (Teleostei: Cottidae), a predator of sea anemones. *J. Exp. Mar. Biol. Ecol.* 204 (1–2): 23–42.
Zaniolo, G., L. Manni, R. Brunetti, and P. Burighel. 1998. Brood pouch differentiation in *Botrylloides violaceus,* a viviparous ascidian (Tunicata). *Invert. Rep. Dev.* 33 (1): 11–23.

Zimmer, Russel L. 1997. Phoronids, brachiopods, and bryozoans. In *Embryology,* ed. Scott Gilbert and Ann Raunio, 279–305. Sunderland, MA: Sinauer Assoc.

Zimmerman, R. C., D. G. Kohrs, and R. S. Alberte. 1996. Top-down impact through a bottom-up mechanism: The effect of limpet grazing on growth, productivity, and carbon allocation of *Zostera marina* L. (eelgrass). *Oecologia* (Berlin) 107: 560–567.

Zupan, John R., and John A. West. 1988. Geographic variation in the life history of *Mastocarpus papillatus* (Rhodophyta). *J. Phycol.* 24: 223–229.

———. 1990. Photosynthetic responses to light and temperature of the heteromorphic marine alga *Mastocarpus papillatus* (Rhodophyta). *J. Phycol.* 26: 232–239.

INDEX

Page numbers in italic denote illustrations.

Design: Barbara Jellow

Composition: Integrated Composition Systems

Text: 10/14 Minion

Display: Arquitectura, Frutiger

Printing and binding: Imago